## Introduc

On the morning of January 14, 2012, t
airport was quiet, except for one man, likely in his forties, who was
enjoying a few drinks with a friend. His clothes, unfit for Chicago's
winter weather, told me that he was on his way to somewhere more
fun than I was.

I don't remember how we started talking, but soon Charlie asked
me what I did for a living. As a stereotypical scholar, I spilled my guts.
I told him I worked on collective learning and that I was running a
lab at MIT. But when I asked him the same question, he mentioned
being a lawyer and changed the subject. I was intrigued. He told
me that with his girlfriend, who worked at an airline, they traveled
on a whim to sunny destinations. Eventually, Charlie warmed up
enough to tell me his story.

After graduating from college in the 1980s, Charlie began practic-
ing law in Florida. But he suddenly lost his job. While he was still
figuring out his next move, he received a call from a friend who
asked him to appear in court the next day to extend one of his cases.
This required the presence of a certified lawyer, but not necessarily the one handling that case. Charlie's quick-witted response was
simple: fifty bucks!

Soon word got out. For fifty bucks Charlie would file a motion
to extend your case. He started getting calls at 4 a.m. saying:
'Charlie . . . I am shitfaced! Please cover for me tomorrow!' His
response was always the same: fifty bucks!

Decades later, Better Call Charlie* had grown into a successful
law firm specializing in 'court-coverage services.' Charlie's firm
not only grew in volume, but in quality. His team knew the case

* www.bettercallcharlie.com

ix

extension processes better than the lawyers handling the cases themselves. Over a few decades, Charlie transformed that serendipitous phone call into a well-defined niche in the vast world of knowledge. A niche that allowed him and his girlfriend to travel the world unincumbered while his little legal engine kept running. It was the American dream. Not the white picket-fence one, but the one where money is not an issue, and the college party never ends.

Down on his luck, Charlie figured out how to become a perfect piece in the puzzle of complementary knowledge. But Charlie's story is not just a fun anecdote. It is a clear illustration of how nuanced and specific knowledge can get. Court-coverage services in the state of Florida are one of the millions of things in which people can specialize. Like a gene in a cell, a brick in a castle, or a funky new letter in an ever-growing alphabet, Charlie's operation became a cog in a machine composed of lawyers, judges, and courtroom clerks. A simple idea, but one showing that knowledge cannot be simply broken down into a few categories. Like the uniqueness of snowflakes and fingerprints, knowledge is one of the more nuanced concepts in our world. It is not a thing, but an infinite alphabet. An ever-growing tapestry of unique ideas, experiences, and received wisdom.

Writing about knowledge is hard, because even though 'knowledge' is a colloquial word, we use it to talk about vastly different things. Most academics equate knowledge to the facts and theories published in academic papers, books, and patents. But this elite definition leaves out the very valid experiences of car mechanics, gardeners, and musicians. Knowledge is not only the result of research focused on uncovering facts or validating theories, it also includes experiences accumulated more haphazardly across many activities.

Still, scholars can provide useful ways to classify knowledge. Consider the basic plot of a murder mystery. Usually, murder stories start at a crime scene, with detectives collecting factual knowledge. This is knowledge about things, or facts, like the position of a bullet hole or the time of a phone call. But since these clues don't explain

*The Infinite Alphabet*

# The Infinite Alphabet

*. . . and the laws of knowledge*

## CÉSAR A. HIDALGO

ALLEN LANE
*an imprint of*
PENGUIN BOOKS

ALLEN LANE

UK | USA | Canada | Ireland | Australia
India | New Zealand | South Africa

Allen Lane is part of the Penguin Random House group of companies
whose addresses can be found at global.penguinrandomhouse.com.

Penguin Random House UK
One Embassy Gardens, 8 Viaduct Gardens, London SW11 7BW

penguin.co.uk

First published in Great Britain by Allen Lane 2025

001

Set in 12/14.75pt Dante MT Std
Typeset by Six Red Marbles UK, Thetford, Norfolk
Printed and bound in Great Britain by Clays Ltd, Elcograf S.p.A.

The authorized representative in the EEA is Penguin Random House Ireland,
Morrison Chambers, 32 Nassau Street, Dublin D02 YH68

A CIP catalogue record for this book is available from the British Library

ISBN: 978-0-241-65567-2

Penguin Random House is committed to a sustainable future
for our business, our readers and our planet. This book is made from
Forest Stewardship Council® certified paper.

To Anna & Iris

# Table of Contents

how a murder was conducted, or why, detectives need to construct a theory connecting all these facts. That theory is an example of conceptual knowledge.* But to get to that theory, detectives often need to gather additional evidence using procedural knowledge. For instance, the knowledge needed to match fingerprints or sequence DNA.†

Another thing that makes writing about knowledge difficult is that these distinctions can depend on the academic field. In education – a field that is deeply concerned with learning – factual, conceptual, and procedural knowledge form the basis of an important framework introduced in 1956 by Benjamin Bloom.[1] The framework provides a hierarchy of aptitudes, where factual knowledge represents a lower level of understanding than knowledge of mechanisms or the ability to carry out a procedure. In fields focused on strategic decision-making, such as theoretical economics, knowledge is sometimes used as a synonym of information. A key example is Friedrich Hayek's famous 1945 paper 'The Use of Knowledge in Society.'[2] There Hayek uses knowledge to describe the information people have *about* the means, wants, and needs of others.‡ For instance, the information a baker uses to decide how much flour to buy or how much bread to bake.§

---

* Krathwohl, David R. 'A Revision of Bloom's Taxonomy: An Overview.' *Theory into Practice* 41, no. 4 (2002): 212–218.

† Factual, conceptual, and procedural knowledge are prevalent in our world. In some languages, we even use different words for them. In Spanish, as well as in other Romance languages, the verbs *conocer* and *saber* are used, respectively, to talk about factual and procedural knowledge. If you want to say that someone knows how to read, you say: '*él sabe leer.*' If you want to say that someone knows your uncle, you say: '*él conoce a mi tío.*' In English, we use the same verb (to know) but add 'how' to describe procedural knowledge or know-how ('he knows how to read').

‡ This is a form of factual knowledge that is also described in some circles as epistemic knowledge.

§ This confusion is common in academic work focused on decision-making, since what is relevant in those fields is how information changes the decisions that someone makes. For example, how a poker player changes their strategy when they know their opponents' cards.

But today, many scholars embrace a notion of knowledge that transcends information, and have also come to understand that knowledge satisfies some key characteristics.[3]

For instance, knowledge – like information – is non-rival, meaning that it can be copied without being destroyed. Think of learning a song. When you learn a song from someone that person doesn't lose their knowledge of that song. This doesn't mean that all forms of knowledge are easy to copy, but we will get to that eventually.

Another key idea is that knowledge can be tacit or explicit.[4] By tacit we mean knowledge that cannot be written down or communicated using words and pictures. Learning requires practice, often in collaboration with other people who already have the knowledge you want to get. If you've ever played music in a band, or worked closely with a mentor or editor, you know what I mean. These are interactions that help transmit tacit knowledge.

But in this book, we are interested mainly in the idea that knowledge is non-interchangeable or non-fungible. That comes from the uniqueness of knowledge exemplified by Charlie's story. His court-coverage services operation cannot be interchanged – for instance – with a law firm focused on intellectual property disputes. We will get to cases in which this non-interchangeability matters, but for now, all we need to know is that this gives this book its name. The idea that knowledge is made of a myriad of non-interchangeable pieces is why I like to say knowledge behaves like an infinite alphabet.*

But how big is this alphabet? Consider a world where knowledge is specific to industries and occupations. That is, a world where a chemist working for a candy factory is not the same as a chemist working for a copper mine. If we consider 1,000 industries and 1,000 occupations, that means 1 million unique combinations. And that is not even considering knowledge that is specific to regions, since

---

* This is certainly metaphorical. It may well be that – in the future – we will able to reduce knowledge to some type of large yet countable periodic table. To the best of my knowledge, I don't think we are there yet (and I have the intuition that we might never get there, as knowledge may be incomplete in a Gödelian sense).

working at a candy factory in China might not be the same as doing it in Paraguay. Certainly, there are industry–occupation pairs that we should not expect to observe, such as a hair stylist working at a pastry shop. But the point still stands. Knowledge may not be technically an infinite alphabet, but it is at least as large as a dictionary.*

This book derives its title from this simple idea. The idea that to understand knowledge we must transcend the temptation of thinking of it as a single thing. We must accept it as an alphabet. This is the key to unlocking a scientific understanding of knowledge that embraces its combinatorial complexity.

So now that I've clarified a few things, I can finally tell you what this book is really about. In short, it is an effort to summarize and organize our scientific understanding of the growth, diffusion, and value of knowledge. It is a book about the principles that govern how knowledge grows, moves, and decays. It is an attempt to use the stories of maverick migrants, eccentric entrepreneurs, and failed cities of knowledge, to condense over a century of academic work into three laws or principles.

1. The Principles of Time, which describes how people, teams, and industries accumulate or lose knowledge.
2. The Principles of Space, which describes how knowledge diffuses across geographies, social networks, and among economic activities.
3. The Principles of Value, which describes how we can understand the value of the knowledge that agglomerates in countries, cities, and organizations.

The first principle covers how knowledge, grows, shrinks, and is replaced by newer knowledge. It is an attempt to summarize decades

---

* People are sometimes surprised about how small professional circles can be, but this is something that is explained – on average – by the non-fungibility of knowledge. For a simple calculation, consider a world with 1 billion workers. With 1,000 industries and 1,000 occupations, that means an average professional circle (people with the same industry and occupation) of about 1,000 people.

of work on learning curves, progress curves, and experience curves. It is a journey that will take us from a typing class in Pittsburgh in the 1910s to the reasons why Netflix and Amazon were able to beat powerful incumbents during the opening act of the twenty-first century.

The second principle explores how knowledge moves across nearby places and similar activities. It is a principle about how knowledge crosses oceans and mountains, but also about knowledge jumping from one industry to another. It is a journey that will take us from the fathers of American manufacturing during the War of Independence to the reshuffling of the Italian and Japanese aircraft industry in the aftermath of the Second World War.

The third principle focuses on the value of knowledge agglomerations. It is a principle trying to score the words of the game of Scrabble that countries and cities play with the infinite alphabet. It is about connecting the knowledge available in a country or city with its ability to generate economic growth that is green and inclusive.

By the end of this book, I hope you will have a clearer understanding of what knowledge is and how it moves. But more importantly, I hope these laws will give you a conceptual edifice that you can use to accommodate every new piece of knowledge about knowledge that you learn.

After a few beers, I looked over Charlie's shoulder and realized my flight was about to board. Charlie and I took out our cellphones and became Facebook friends. To this day, I see pictures of him skydiving, walking on hot coals, or wearing a shamrock-patterned suit to court on St. Patrick's Day. As I was about to leave, Charlie asked me for a stock tip. I told him I liked Netflix, which had been recently battered for attempting to separate their streaming service from their mail-in DVDs. I never knew if he took the tip, but if he did, he would be up more than 1,000 percent.

# PART I

## *The Principles of Time*

# I.

# *Yachay*

About 120 km north of Quito, on what was until recently an active sugarcane plantation, you will find Yachay.

Yachay means knowledge in Kichwa, the native language of the Incas. But to us, it is the name of a project involving the construction of a city of knowledge and a technical university. The idea was for the city and the university to work as a flywheel, with the university providing the talent for the city's jobs.

The construction of Yachay began with a massive commitment. The original budget was about 1 billion dollars, a considerable amount for a country with a GDP of only about $100 billion. The agricultural lot chosen for the project was also large, about the size of Manhattan, and the investment largely followed through. An official report looking at the first eight years estimated $600 million in expenses.[5] Other estimates put the cost of the project at $1.2 billion.[6]

Yachay was inaugurated in 2014 by the then president of Ecuador, Rafael Correa.[7] But he was not the only one enthusiastic about the project. José Andrade, a professor of engineering at the California Institute of Technology (Caltech) declared that he was 'personally in love with [Yachay]. It's one of the greatest things that I've seen in this country, ever.'[8] Fernando Cornejo, Yachay's academic manager, was also ecstatic. He proclaimed that finally: 'there will be no more crowding! Each classroom will only have fifteen students.'[9]

But that was the problem. There was no crowding.

In only a few years the promise of Yachay began to unravel. Building a university and city of knowledge in the middle of nowhere is expensive. The lack of proper infrastructure meant even water had to be brought regularly on cistern trucks. What started as a romantic utopia devolved into an organizational nightmare.

Paola Ayala moved to Yachay in 2015 from the University of Vienna. She had completed a PhD in physics and wanted to 'help change [her] country.' She was fired in 2017 together with five other scientists, including the chancellor and a geologist recruited from East Carolina University.[10]

People began to turn on each other. The university's president, Carlos Castillo-Chavez, derided the faculty's research output. The faculty pushed back, accusing Castillo-Chavez of steering the project away from a research powerhouse and into a run-of-the-mill teaching university.

But Yachay is not alone. It's just one of the many stops you'll find on the boulevard of broken development dreams.*

Neom is a more recent stop on that boulevard. A planned smart city in the northwest of Saudi Arabia with an estimated cost of $500 billion.[12] Neom's plans are straight out of science fiction, with features like The Line, a car-free city for 9 million people stretching for 170 km (110 miles). But despite the plush budget, the planned smart city is already encountering problems. Foreign consultants, sometimes offered tax-free salaries in the range of $700,000 to $900,000, have a colorful way to describe their experience. 'When you start at Neom, you bring two buckets. The first is to hold all the gold you'll accumulate, and with so many living expenses taken care of, it will soon grow heavy. The second bucket is for all the shit you take. When that bucket is full, you pick up your bucket of gold and leave.'[13]

Yachay and Neom teach us valuable lessons. On the one hand, they tell us that governments around the world care about growing the knowledge base of their economies. This is an upright

---

* This sentence is inspired by Joshua Lerner's book *Boulevard of Broken Dreams*, which focuses on public efforts to support entrepreneurship.[11]

intention. But on the other hand, these projects tell us that governments around the world do not know how to grow knowledge. The projects are built with a mentality similar to that of a famous quote from *Wayne's World 2*: 'If you book them, they will come.'* But what worked for Wayne doesn't seem to work in reality. So, to understand where Yachay and Neom went wrong we need to look beyond intentions. We need to understand what the forces are that shape the growth, movement, and value of knowledge.

The idea that knowledge is at the core of the wealth of nations became a staple of economic theory more than thirty years ago. Economists working on what we now know as 'endogenous growth theory' showed that economic growth must come from something that is non-rival.† Something that can be copied like a song, not shared like a hammer. Today, people working in international development agencies, government, and academia, are more likely than not to know the name of Paul Romer, the 2018 Nobel Prize-winning economist who, among others, helped identify knowledge as the secret to the wealth of nations. Paul was part of a cohort of economists that included Philippe Aghion, Peter Howitt, and Paul's own PhD advisor, Robert Lucas, who transformed the idea of the 'knowledge economy' into a global phenomenon.[14,15] A key contribution to this transformation was a mathematical model Paul published in 1990. The model incorporated a simple but powerful idea: the notion that the non-rival nature of knowledge was needed to explain economic growth.

For the better part of the twentieth century, models of economic growth assumed that economies employed two factors: capital and labor. You can think of labor as people, and capital as the tools that people use to build things. Imagine a team of carpenters using nails, boards, and hammers to build birdhouses. Together, capital and labor produce output. But they are constrained by the fact that

---

* A quote that is a parody from *Field of Dreams*: 'If you build it, he will come.'
† Something that is not destroyed when copied or shared.

labor and capital are *rival* inputs: only one person can use a hammer at any point in time.

In these models, economies grow by saving a fraction of what they produce and investing it in new capital (e.g. new hammers). This transforms savings into the capital they need to grow.

The problem with these models is that rivalry carries an important limitation. In a world where ten carpenters use ten hammers to produce ten birdhouses, producing twenty birdhouses requires twenty carpenters with twenty hammers. In that world, even though the output is increasing, the output per unit of capital (hammers) and labor (carpenters) is not. A world made of only rival inputs struggles to explain economic growth.

The answer brought by Romer and his contemporaries was rather intuitive. Growth must be the consequence of economies accumulating a non-rival input, something like ideas, information, or knowledge. Since these inputs can be copied, they can grow in per-capita terms.

Imagine a carpenter who learns how to drive a nail with a single strike. That carpenter can tell others about his technique without losing it. As other carpenters learn the technique, they produce birdhouses faster. Now imagine a carpenter who invents a nail gun, a tool that embodies knowledge. As these improvements percolate through the system, ten carpenters learn to produce more than ten birdhouses. Economic growth comes from the accumulation of knowledge.

This simple idea was revolutionary. It not only catapulted Romer to global fame but changed the global conversation on economic development. Knowledge became a central concern for international development efforts, which began demanding technical assistance and capacity building components. So, we cannot blame Ecuador or Saudi Arabia for betting on knowledge. We all agree that knowledge is the goal. But unfortunately, the world is not as simple as carpenters and hammers.

Knowledge can be copied, but that doesn't mean that it is easy to copy. Music provides a great example. Buying a guitar is much easier than learning how to play it. Of course, there are simple forms of factual knowledge that diffuse easily, like knowing that Madrid is the capital of Spain. But the world of knowledge is not just a collection of simple facts. The knowledge needed to discover and test a gene therapy, build a commercially viable excavator, or succeed in the race for the next generation of semi-conductors involves countless bits of procedural, conceptual, and factual knowledge that are hard to accumulate and transfer. Knowledge is non-rival, but it is also non-fungible. The first property makes it copiable. The second one makes copying it like trying to move a 1,000-piece jigsaw puzzle with your bare hands from one table to the next.

So where do moonshots like Yachay and Neom often go wrong?

Knowledge is certainly a valid goal. But knowing what the target is and knowing how to reach it are two very different things. Humans have been able to estimate the distance to the Moon for more than 2,000 years, but we only recently figured out how to get there. Just as building rockets requires knowing about physics efforts to engineer the growth of knowledge must rely on some scientific principles. Principles explaining the growth and diffusion of knowledge.

We need these principles because growing knowledge is a rather delicate process, and one that is also different than developing capital-intensive megaprojects. Consider the construction of oil refineries, steel mills, and power plants. These are often centered on a key piece of infrastructure and require accumulating knowledge on a relatively narrow domain. They focus on a known industrial process that has been replicated numerous times. Cities of knowledge, like Shenzhen, Paris, or Boston, are very different from a capital-intensive megaproject. On the one hand, infrastructure, from research buildings to meeting rooms, plays a much more muted role in a city of knowledge than it does in a monolithic industrial activity like a steel mill. On the other hand, remoteness, which can be an asset for the construction of a power plant, represents a challenge for a city of knowledge. Innovation concentrates in cities,[16,17] not by design but by necessity, since knowledge creation requires frequent interactions among people with complementary skills.[18,19] These human networks are the true infrastructure that underlies knowledge accumulation, but unlike buildings, they are much harder to engineer. Social and professional networks often grow slowly and organically as they accumulate knowledge that becomes hard to transfer. The buildings are the epiphenomenon of the city, not the other way around.

The studies showing that knowledge is hard to move are contemporary to the work of Romer.[16,20–22] These are studies of spillovers connecting inventors and scholars using data on patents and publications. They are part of an empirical literature that, while important, is not as popular as endogenous growth theory. After all, Romer's ideas are extremely hopeful. They tell us that knowledge can be copied, and that knowledge is all we need. The empirical literature is grimmer. It tells us that copying knowledge is difficult, and that knowledge tends to concentrate in a few places. A naïve reading of Romer tells us that knowledge can potentially grow anywhere. The empirical literature tells us that knowledge is a source of spatial inequality. Unlike a drop of ink, which diffuses effortlessly in a glass of water, knowledge behaves more anti-diffusively, agglomerating in cities while constrained by pre-existing social networks, language barriers, and by knowledge's own complementarities.

Attempts to engineer cities of knowledge like Yachay or Neom seem to ignore this behavior. They largely focus on physical infrastructure in remote locations. The results are soulless architectures that make starchitects rich but are unlikely to generate the networks needed to grow knowledge. The best chances for Ecuador to grow its knowledge economy are in Quito or Guayaquil, not in the unpopulated highlands of Urcuqui. I remember trying to communicate this to a delegation from Yachay that visited my lab at the Massachusetts Institute of Technology (MIT) in 2014. The project was only getting started, but I could see where it was going. The small team of delegates, however, were not looking for feedback. They were convinced that a billion dollars and an agricultural plot in the high Andes were all they needed to compete with Silicon Valley.

But by setting up in a remote location and focusing on infrastructure they were making the difficult impossible. After all, Apple, Google, and Hewlett-Packard did not need fancy buildings to get started. They began in modest garages, just as Netflix and Microsoft started in strip malls.[23,24] Physical infrastructure was not the constraint. Our understanding of how knowledge grows and flows still might be rudimentary, but it is good enough to understand what pitfalls we should avoid. It tells us that knowledge cannot be engineered as easily as a power plant. It tells us that the failure of Yachay is not the opposite of the success of Charlie, since Charlie succeeded by discovering a key specialization in an ecosystem that was already rich in complementary knowledge. Yachay was an attempt to grow a full ecosystem of knowledge from scratch.

In the future, if we are ever able to engineer a knowledge economy, it will be through efforts that probably won't look like Neom or Yachay.* Later in this book, we will see some efforts that, while contemporary to Yachay, did things very differently. We will look at the story of China's 'Silicon Valley,' the Beijing neighborhood of Zhongguancun.†

On January 6, 2023, the president of Ecuador Guillermo Lasso signed executive order 639, closing the state-owned enterprise in charge of the 'City of Knowledge.'[26,27] The university, Yachay Tech, will continue its operations. And for what it's worth, despite all its problems, it appears to have produced a few cohorts of motivated

---

* During the time I was writing this book, Saudi Arabia announced important cuts to The Line, a key component of the city of Neom. The Line, which was expected to cover a 170-km stretch along the coast, was pulled backed to just 2.4 km and expected to be completed by 2030.[25]
† We will also look at the story of the Universidad San Francisco de Quito, which is by many accounts the most succesful university in Ecuador.

students. Four hundred and fifty students, to be precise.[5] But at what cost? After over $600 million in investment the university has twenty-two classrooms, with a combined capacity of less than 1,500. The future of Yachay looks uncertain as the remote university learns to live within its means.[5]

## 2.

# The Curve

There is a beautiful documentary you may want to watch one day: *A Band Called Death*. The documentary tells the story of an American rock band from Detroit, Michigan, which, despite being largely unknown, invented the music genre we today recognize as punk. Formed in 1971, well before the Ramones and the Sex Pistols, the Hackney brothers developed the sound of a rowdy and disenfranchised youth. But the Hackney brothers did not look like your stereotypical punk trio. They did not spike their hair with gel or wore jackets with metal studs. Instead, they rocked Afros and wore regular American sportswear.

Sometimes, stories don't begin where we think. And in our case, the quantitative history of knowledge does not start with Paul Romer, or even with an economist. Decades before Romer was born a psychologist called Louis Leon Thurstone took some of the first steps.

Thurstone* began his career as an engineer and an inventor. But it was hard for him to stick to a single discipline. His father had also wandered across many professions, starting out as a Swedish Army mathematician and later becoming a Lutheran minister, a newspaper editor, and a publisher. Louis began his career by studying electrical engineering at Cornell. He designed a motion picture camera that impressed Thomas Alva Edison enough for

* A thorough biography of Louis Thurstone can be found in this national academy of sciences biographical memoir by J. P. Guilford: www.nasonline.org/publications/biographical-memoirs/memoir-pdfs/thurstone-louis.pdf

him to offer Thurstone a job. But Thurstone didn't stay for long with the Wizard of Menlo Park. He soon moved to the University of Minnesota to teach engineering and geometry. It was there that his light bulb switched on.

The struggle to teach students made Thurstone curious about learning, so he decided to – once again – switch careers and enroll in the graduate school of education at the University of Chicago. After a brief period, he transferred to the psychology PhD program where he would complete his doctorate,* 'The Learning Curve Equation.'

His thesis leveraged data from a typing class conducted in 1916 at Pittsburgh's Duff School of Business. Every week, students took a four-minute test intended to assess their skills. Thurstone used these tests as data to model how fifty-one seventeen-year-olds learned to type.†

Today typing seems easy. Children grow up with their parents' phones in their hands. But in the 1910s typing was something that few people had tried. This meant Louis could measure learning from its onset, as students gathered speed and became confident typists. At the beginning, students typed as little as twenty-seven words in four minutes,‡ but they improved weekly. The fact that people learned was of course obvious enough, but what was not obvious was the shape of the mathematical function governing learning. Thurstone found that learning followed a well-defined

---

* see www.researchgate.net/publication/316981670_Thurstone_LL
† The class contained eighty-three students, but he kept data for fifty-one, since many students dropped out or attended irregularly.
‡ Thurstone's full thesis can be found in the Internet Archive at: archive.org/stream/learningcurveequoothurrich/learningcurveequoothurrich_djvu.txt
'I asked ten of my students who had never touched a typewriter to take a four-minute test. The average score for this group was twenty-seven words in four minutes and this is used with the other data as an average initial score in typewriting.'

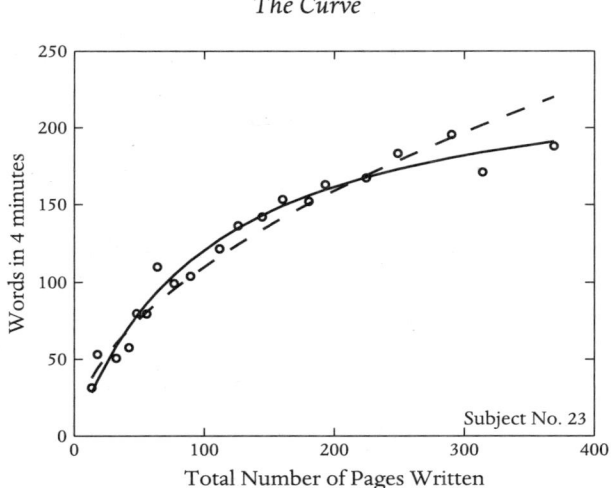

*Figure 1: Learning curve for Subject No. 23 in Thurstone's PhD thesis. Circles show experimentally recorded values. The solid line shows Thurstone's learning curve model and the dotted line a power-law fit (y~x^a).*

mathematical form,* a statistical law representing an early example of a learning curve.†

Thurstone's function tells us that the rate at which students learn decreases with practice. At the beginning, learning is 'steep.' The first hundred pages brought students from about twenty-five words in four minutes to more than a hundred. But as they continued to practice, their rate of improvement petered out. It would take ten weeks – on average – for students to reach twenty words a minute,

---

* Thurstone found that the learning rate of students followed a function of the form $y \sim L(x+p)/(x+k)$, where $y$ is the amount of time it took a student to type a number of words (the students' performance), $x$ is the total amount of practice the student had (measured as the number of pages the student had already typed), and $p$ and $k$ were parameters modulating the relationship between experience and performance.

† An earlier version of the learning curve involves the work of the German psychologist Hermann Ebbinghaus, who also studied forgetting curves in his 1885 book *Memory: A Contribution to Experimental Psychology* (Teachers College, Columbia University, New York, N.Y., 1913).

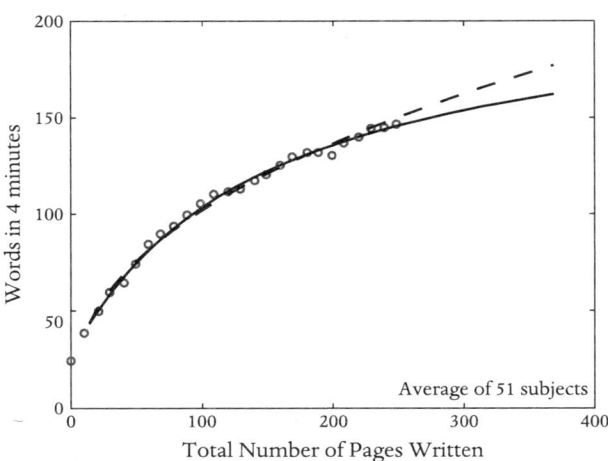

*Figure 2: Average learning curve for Thurstone's typing class students. Circles show experimental values. The solid line shows Thurstone's learning curve model and the dotted line a power-law fit* ($y \sim x^a$).

and they would only be able to double that speed after sixteen more weeks. Eventually, learning was bound by a natural limit. A parameter that Thurstone called $L$ representing the maximum number of words that an average student could type in four minutes.

Using his data Thurstone estimated $L$ to be around fifty-four words a minute. A simple Google search reveals that today people type at a speed of about forty words a minute and that experts type at about twice that rate. So Thurstone's estimate of fifty-four is compatible with that of a skilled modern writer.★

But what is important for us, is that Thurstone's work showed that learning could be treated as a quantitative phenomenon, providing a point of departure for a future research. His experiments are peppered with clever observations, such as the idea that learning is not a function of time, but experience. If we pay close attention

★ These figures reproduce, respectively, figure 3 and figure 5 from Thurstone's PhD thesis. Solid lines show the fits provided in his thesis. These are respectively, $Y=244X/(102+X)$ for figure 3 and $Y=216(X+19)/(X+148)$ for figure 5. Dashed lines show power-law fits. These are respectively $Y=9.56X^{0.53}$ and $Y=14.63X^{0.42}$.

to his charts, we will see that the $x$-axis is not measured in days but in number of pages written. But Thurstone's work is only the beginning. A few decades later, the same idea will reemerge not far from Pittsburgh, thanks to another engineer working at the heart of America's aircraft manufacturing industry. His name was Theodore Wright.

Theodore Paul Wright was born eight years after Thurstone, in 1885 in Galesburg, Illinois.[28] He was the son of a prominent Massachusetts family headed by Philip Green Wright, an American economist and instructor at Harvard University known for proposing instrumental variables as a method to pinpoint causal effects. Theodore's older brother Sewall probably became the most famous member of the Wright family – he was an evolutionary theorist with a PhD from Harvard University who together with Ronald Fisher and J.B.S Haldane helped formalize Darwin's theory of evolution.

But Theodore had more than enough merits of his own. In 1918, he received a bachelor's degree from MIT and soon after he began working as an aircraft inspector. In 1922, he moved to Curtiss Aeroplane and Motor Company, and in 1925 he was made chief engineer.

Wright lived through the golden age of aviation. He supervised the creation of race-winning planes with cool names such as Condor and Hell Diver. But he was not only an aircraft engineer, he was a key player in the development of an airplane manufacturing system that became an industry standard. In the 1930s, he was at the top of the aeronautics world, running the Curtiss-Wright corporation, first as vice-president and general manager, and later as director of engineering at the parent company.

Wright's genius didn't go unnoticed. In the 1940s the government named him director of the office responsible for aircraft production, and President Roosevelt called on him to produce 50,000 planes a year. Wright doubled that goal.[28]

But why was he so confident? Throughout his career Theodore had studied how teams became better at building planes. In 1936, he carved some time out of his busy schedule to publish a paper called

'Factors Affecting the Cost of Airplanes.'[29] As you may anticipate, his discovery was eerily similar to that of Thurstone, even though he doesn't cite the psychologist's work.

Wright showed that the cost of a plane, what in Thurstone's experiment would be the time needed for a student to type a word, decreased with experience. Not only that, but that the cost of producing a plane decreased as a power-law, a mathematical formula that also fitted Thurstone's data.

Wright used the fact that planes are produced in batches to study the cost of the last plane in a batch. The cost of the last plane in a ten-plane batch was about half that of a plane made in a batch of one. And like Thurstone, Wright also found improvements petered out, with most of the learning accumulating at the beginning of each batch. In a twenty-plane batch the cost was about 41 percent of the original cost, and in a thirty-plane batch, the cost dropped only a few additional percentage points, to about 38 percent.*

Thurstone and Wright discovered the same principle. A principle governing the growth of knowledge with experience. Even though the context was quite different, the similarity of their results suggested that something fundamental was at play. In the case of Thurstone, learning had a clear interpretation. After all, what else could have changed? The same students were using the same typewriter week after week. They were producing more with the same, fitting squarely within Romer's theory. But in the case of Wright's study, learning was only one of many possible interpretations. Maybe, what we were observing was not learning, but improvements in technology? Maybe tools were getting better? Or maybe teams were becoming larger? Moreover, plane manufacturing could be benefiting from economies of scale, with time and cost reductions explained by cutting metal in batches. These are all valid

---

* Another noteworthy example here is the work of Harold Asher, who produced a similar study a couple of decades after Wright: Asher, Harold. 'Cost-quantity Relationships in the Airframe Industry.' PhD diss., Ohio State University, 1956.

criticisms of Wright's findings, which were addressed three decades later by an economist named Leonard Rapping.

Rapping was born in Indianapolis in 1934, a few years before Wright published his landmark paper. For his studies, he traveled west to complete an undergraduate degree at UCLA. He then returned to the Midwest for his doctorate, joining the University of Chicago where he became famous among economists for his work with Robert (Bob) Lucas on business cycles. Like Wright, Rapping was publicly minded and participated in numerous government agencies and committees, from the Federal Bureau of the Budget to the Defense Department.[30]

But Rapping also had a soft spot for the study of learning. In his case, he took advantage of shipbuilding data collected during the Second World War to explore the increase in productivity observed across multiple shipyards. Rapping observed that between 1941 and 1944 the productivity of shipbuilders had grown at a rate of about 40 percent a year.[31] This was huge compared to the average rate of the US economy, which has historically grown at about 2–3 percent a year.

Rapping's study focused on Liberty ships; a relatively inexpensive service ship that was mass produced across America. He used the fact that some shipyards were established before others to isolate the effects of learning. Technology, after all, was improving everywhere, but experience could only accumulate in a shipyard after it started operating. This allowed Rapping to control for some of the factors that were unaccounted for by Wright. His conclusion, however, was the same. The increase in productivity observed in these shipyards was not coming from economies of scale or changes in technology. In his own words: 'the secular improvement in output per man-hour during the war is attributable to learning.'[31]*

---

* Another hypothesis suggested by my editor was that this could be due to patriotism, which might have made people work harder. Yet, it would be hard to argue that patriotism was increasing during the period (workers might have

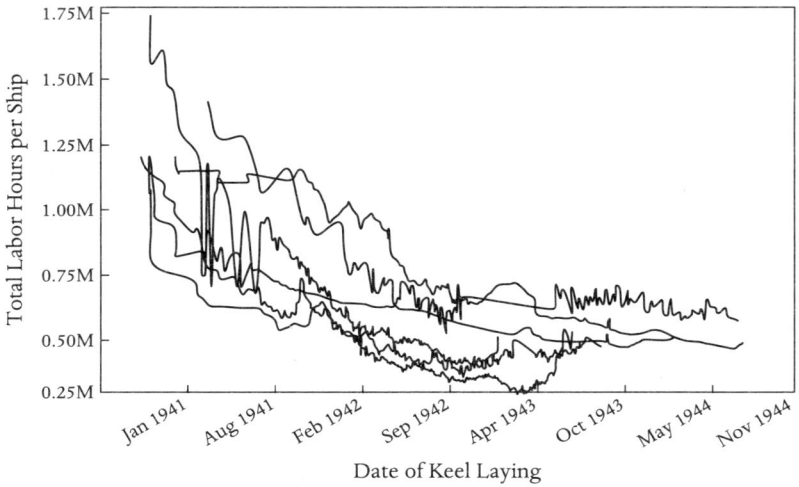

*Figure 3: Days needed to complete a Liberty ship. From Peter Thompson, 'How Much Did the Liberty Shipbuilders Learn? New Evidence for an Old Case Study,'* Journal of Political Economy *(2001).*[32]

The learning rates observed in the manufacturing of Liberty ships were nothing short of extraordinary. Peter Thompson, an economist now at Georgia Tech, showed that while the first ships produced in each yard required more than sixth months to be completed, only thirty days were required to complete a ship at the end of 1943.[32]

Thurstone, Wright, and Rapping provide a powerful point of departure for the study of learning. Their findings are consistent and valid across different scales, from lonely typists to shipbuilding teams. But what may surprise you is that this principle of learning also applies to learning machines.

In recent years, several companies, such as Google and OpenAI, have released powerful artificial intelligence models that are becoming increasingly better at generating language and images. Like

---

been equally patriotic in 1943 and 1944). Also, it is implausible that this increase in patriotism started at different points in time in ways that coincided with the manufacturing of a each shipyards' first ship.

Thurstone's typing class students, these models are not born ready. Models also 'go to school' by looking at millions of examples of images and text. Guess what is the shape of the curve describing their performance as they swallow increasingly more data? If you guessed the same power-laws described by Thurstone, Wright, and Rapping, you would be correct.

A recent study by a team of OpenAI researchers shows this for a language model, a 'relative' of today's famous ChatGPT.[33] The researchers studied the 'test loss,' a measure of the model's accuracy, as a function of the data used to train it. They found the loss decreased as a power of the size of a dataset, echoing the work of Thurstone, Wright, and Rapping. In this example, however, learning was rather slow, with test losses dropping only by about 5 to 10 percent with each doubling of the data, about half the typical learning rate observed in manufacturing.[34] But this is not an isolated case. A similar study by a team at Google also found learning curves like those of Thurstone and Wright when studying how computer vision models learn as they 'study' more images.[35]*

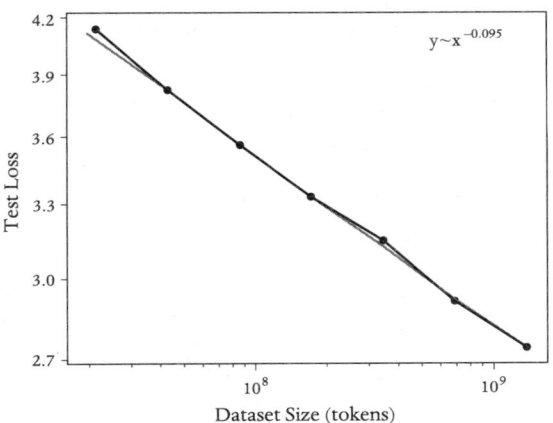

*Figure 4: Model performance (lower test loss means higher performance) of a language model as a function of the size of the dataset. From Kaplan et al.[33]*

---

* Both teams also found larger models learned faster and reached an overall better final accuracy when trained using the same amount of data.

Unfortunately, neither of these studies, nor much of the literature on AI, has connected machine learning with the long literature on learning curves. Still, it is profound to think that the curves describing modern machine-learning models resemble those drawn over a century ago by a psychology PhD student studying a typing class.

This apparent coincidence may run deep, since both teams and machines learn by rewiring networks and are constrained by attention. It is by now common knowledge that machine-learning models learn by adjusting the weights of links in multilayered networks. These weights help encode, for instance, which words in a sentence a model must focus on to determine the context of another word. This is what computer scientists call 'attention.' Organizational learning also involves processes that are constrained by attention and that involve the rewiring of networks.

The idea that organizational learning is constrained by attention dates back to the influential 1958 book *Organizations* by Herbert Simon and James March.[36] A key point in that book is that, since organizations cannot pay attention to everything, understanding their decisions requires understanding how they allocate their attention. But organizational learning also involves changes in networks connecting people, tools, and ideas. This is the basis of an organizational learning model proposed by Linda Argote, a professor at the Tepper School of Business at Carnegie Mellon University (CMU).

Argote likes to think that some of the knowledge in an organization is contained in a network connecting people, tools, and tasks (which include rules, procedures, etc.).[19,37] For instance, the network of people and tools stores knowledge about who knows how to use which tool. The network connecting people and tasks stores knowledge about who is good at each task. In total, there are nine interdependent networks, from the people–people network – storing social and professional interactions – to the tool–tool network – connecting tools that are used together in a task. In this framework, organizations learn by rewiring these networks – by discovering which tools are better for which tasks, which people are better at using each tool, and which tools go well together. As they adapt their interactions, organizations increase their performance, climbing up the learning curve. Just as neural networks learn by adapting the weights among their many nodes, organizations learn by rewiring their internal networks. This is a simple but important point, because it tells us that organizational knowledge cannot be reduced to the knowledge of its individuals. There is knowledge that is stored in interactions.

From an otherwise unremarkable typing class in Pittsburgh, to the latest transformer models, there seems to be some universality to learning – a curve that describes how shipbuilders, aircraft manufacturers, and 'thinking' machines learn. Yet, this simple principle leaves many key questions unanswered.

For starters, it is a principle that involves learning within a single domain. Typists becoming better at typing, shipbuilders becoming better at shipbuilding, and image recognition algorithms becoming better at recognizing images. But how does the knowledge we accumulate in one domain transfer to the next? How do shipbuilders benefit from the experience of airframe manufacturers? And if they do, how does this knowledge get around? More poignantly, there is something unsatisfying about a principle of knowledge predicting that learning always peters out. We can agree that this may be the case for airframe manufacturers, shipbuilders, and a single machine-learning model, but it is a fact that seems counterintuitive in a world where progress has been accumulating for centuries. Also, this principle doesn't tell us much about forgetting. Are there laws that govern forgetting too? Moreover, learning curves tell us little about disruptive innovations or about the geography of knowledge. Airplanes are clearly not the result of gradual improvements in kite manufacturing, and modern machine-learning technologies seem to have erupted only in a few places. There must be something else going on.

As the ink on Rapping's paper was starting to dry, an engineer published another empirical law that did not exactly match the learning curves observed in shipyards and hangars. The engineer would become world famous as one of the founders of the most iconic companies of the digital age. His name was Gordon Moore.

# 3.

# *Moore Was Different*

During the 1940s, Bell Labs was looking for a technology to improve the speed of telephone switchboards.[38] Back then, routing telephone calls required the use of complex mechanical gears, which companies preferred over vacuum tubes because they consumed less power. But gears were becoming increasingly impractical with the explosive growth of telephone lines. That meant researchers at Bell Labs were looking for a low-power, compact, and reliable alternative. A small device that consumed little power and could switch faster than a gear: the transistor, a switch and amplifier the size of a pencil eraser was the answer to that call.*

Today transistors are everywhere. They are probably the best

---

* We all have the intuitive sense that transistors are the cornerstone of our digital society. But what is a transistor? To understand how a transistor works I will use a river analogy. This is not perfect, but it can give us a basic intuition. Imagine a river running down a valley. If the river encounters a cliff, it encounters what in electronics we know as a diode: a one-way valve. Rivers can only flow down a cliff. Diodes are an important precursor of transistors, but they are unable to act as a switch or an amplifier. They just make sure current can only run in one direction. Now imagine a river encountering two cliffs, forming a ravine or canyon. If the space between the cliffs is empty the water will fall into the canyon and the river will fail to flow across it. The switch is off. But if the canyon is already full of water, and the canyon is now a lake, the river will be able to flow across it, assuming there is an exit on the other side of the lake. The switch in that case is on. That's basically how a transistor works as an on-and-off switch: by filling up or emptying a 'ravine,' not with water, but with an electrical charge. Now imagine you can control that lake's water level. If you stuck a giant hose at the bottom of the lake to move the water level up and down, the flow of the river going across will pick up that pattern. The flow will increase when the lake is full

example of how explosive the growth of knowledge can be. But transistors were not everywhere in the 1950s. In fact, back then a single military-grade transistor could cost about $100 (~$1,000 in today's money). This means that, at 1950s prices, the chip inside your phone would cost about a trillion dollars.

This growth also impacted the time it takes to manufacture a transistor. In 1954, the United States produced a grand total 1.3 million transistors,[38] meaning that producing the 16 billion transistors in an iPhone 14 would have taken 12,000 years. That type of growth cannot be explained by the learning curves of Thurstone, Wright, or Rapping. It would be as if one of Thurstone's typing class students went from seven words a minute to about 1 million. That would be impossible, even with sixty years of practice. So, in the world of transistors, something else must be going on. Something that has allowed the number of transistors crammed into a microchip to break Thurstone's ceiling.

Transistor technology follows what in the technical literature we call an experience curve. Unlike the learning curves of Thurstone, experience curves grow exponentially. That means that they do not peter out but continue to rise at an accelerating rate. The best-known example of an experience curve is the one describing the evolution of the number of components in an integrated circuit. In 1965, the same year that Rapping published his learning curve paper, Gordon Moore, the legendary American chemist and Intel co-founder, published his eponymous law.[39] Moore's law states that the number of components in an integrated circuit – such as transistors – doubles every eighteen months. Eventually, Moore's law was revised to a doubling every two years. The point, however, is that there is a second curve describing the growth of knowledge that behaves quite differently from the curves discovered by Thurstone and Wright. A curve where growth seems unbounded. In this chapter we will reconcile these two laws by looking into the history

---

and will dwindle as the lake grows emptier. In this case, the transistor would be working as an amplifier.

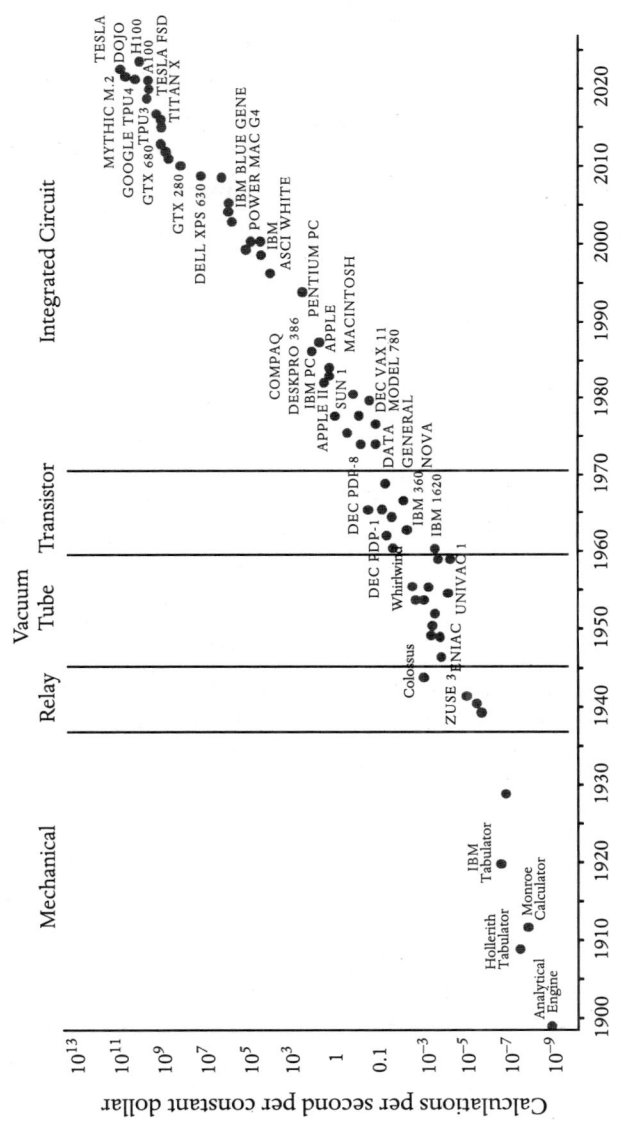

Figure 5: Moore's law. License cc by: www.flickr.com/photos/jurvetson/51391518506

of the transistor. As we will see, this is not the story of a lonely genius, but that of a tumultuous journey through an environment divided by competition and cooperation. It is the story of knowledge growing beyond the boundaries of the firm.

Moore's story begins at the Cosmos Club in Washington, D.C.* He was attending a presentation on transistor physics by William Shockley, a physicist who had co-invented the transistor together with John Bardeen and Walter H. Brattain at Bell Labs. Their 1948 breakthrough made Shockley a scientific celebrity. And he carried that honor with flair. At the Cosmos Club Shockley ended his presentation by tossing transistors into the audience. Moore was impressed.

In 1954 Moore was a young chemist with a bachelor's degree from UC Berkeley and a PhD from Caltech. He had recently moved to Maryland to work at Silver Spring's Applied Physics Laboratory (APL), driving across the country with his wife Betty in a 1950 Buick. But Shockley was moving in the opposite direction. After a long period working in the east, he was heading back west. Worried about being left out of an industry he helped start, he decided to take a leave of absence from Bell Labs and join Caltech as a visiting professor. In California, he struck up an intellectual romance with Arnold Beckman, an experienced American chemist, inventor, and entrepreneur. Beckman had invented the world's first successful electronic pH meter in a collaboration with Sunkist Growers, a large American citrus firm.[40] This motivated him to found Beckman Instruments, a company that continues to operate to this day.

Shockley was able to persuade Beckman to support him by investing in a new venture: the Shockley Semiconductor Laboratory at Beckman Instruments. The lab would be located in Mountain View, California, a concession to Shockley's childhood connections to the Bay Area, and would focus on producing silicon transistors

---

* This section builds on the book: Thackray, A., Brock, D. C. & Jones, R. *Moore's Law: The Life of Gordon Moore, Silicon Valley's Quiet Revolutionary* (Basic Books, 2015).

at a time when most transistors were made of germanium. Even though germanium transistors switched faster than silicon they are less resistant to heat, giving silicon an advantage for military use.*

Shockley needed to hit the ground running. In 1955, the United States manufactured over 3.6 million germanium transistors and over 90 thousand silicon transistors.† Texas Instruments, who recruited Gordon Teal out of Bell Labs, was becoming a leading transistor producer, releasing the first transistor radio (the TR-1)‡ in October of 1954.§

There was also significant pressure coming from outside the United States. In 1948, two German physicists, Herbert Mataré and Heinrich Welker, had developed a germanium point-contact transistor in Paris,[41] which they bizarrely named the 'transistron.' The Soviet Union was not far behind. They manufactured their first transistor in 1949 and established a dedicated institute for semiconductors in 1953.[42] Transistor technology was mushrooming around the planet and Shockley needed to get a firm foot in the race.

Both Beckman and Shockley believed in the power of individual genius, so they began looking for the best people they could find. In February 1956, Shockley phoned Moore, who after visiting Shockley's Mountain View operation agreed to join him in California. Moore also had family ties to Silicon Valley. He grew up in Pescadero and in 1950 married Betty Whitaker from Los Gatos, who became his lifelong partner. Moore agreed to join Shockley for a salary of $750 a month (about $9,000 in 2024) and Shockley asked Moore to leave his job at APL no later than May 1.

Shockley's scientific acumen, however, did not translate into managerial skill. He organized his semiconductor company as a

---

* This was particularly the case in California, in the aftermath of a war that caused the military to expand considerably toward the west.
† Shockley had not manufactured any of them.
‡ The TR-1 was advertised as measuring 3"×5"×1.25" compared to Sony's TR-63 which was 4.4"×2.8"×1.3" .
§ In a collaboration with Regency Electronics.

research laboratory, with some people working with him on academic papers and others, such as Moore, on the transistor project. During 1956, Moore dedicated himself fully to making transistors, building the equipment required to manufacture them using his glassblowing skills. But despite Moore's best efforts, by the end of the year they were nowhere near a shippable product. Shockley, however, was living the dream. He received the 1956 Nobel Prize in Physics together with John Bardeen and Walter Brattain and spent the early months of 1957 touring Europe and the United States. Back in California, problems began to grow.

On the one hand, Beckman was getting worried. He had originally agreed to fund the venture for two years, and eighteen months in, he needed a reason to keep going. On the other hand, Shockley was spending more money than they had originally agreed to. Moreover, he was moving people out of the transistor project after growing obsessed with what he thought was the next big thing: the four-layer diode. Moore, together with Robert (Bob) Noyce – a brilliant physicist who would later join Moore as a co-founder of Intel – wrote Shockley a memo urging him to keep focus on the transistor. But the Nobel Prize-winner was unmoved.

In May 1957, Beckman visited Mountain View and asked Shockley to reduce costs. The physicist exploded in front of everyone. He told Beckman that if he did not like what he was doing he would leave with his entire team. That turned out to be Shockley's key mistake.

After the meeting, Moore, Noyce, and other engineers met for lunch and decided that Moore would tell Beckman that the team was not eager to leave with Shockley. This piqued Beckman's interest, who began flying more frequently to the Bay Area. On June 1, he told Shockley that he wanted to introduce a change in management. Shockley would retain the scientific direction of the lab but would rely on a professional manager for daily operations. But when the team learned about the changes, they felt Beckman had taken Shockley's side. Both bridges had been burned.

Even though Moore and the others could easily find other jobs, they understood that they were more valuable together. At a meeting at Moore's house, Eugene Kleiner – another member of the transistor team – mentioned that his father had contacts in the finance industry. So, they asked him to pen a letter* to his dad asking him for a corporation interested in seven scientists and engineers promising to deliver a diffused silicon transistor† within a year.

That letter caught the attention of Alfred Coyle and Arthur Rock, respectively a partner and a young analyst at Hayden Stone.‡ Arthur would later become one of the most celebrated venture capitalists in the history of Silicon Valley. By August, Coyle and Rock had managed to capture the interest of Sherman Fairchild, a son of one of the original investors in IBM. Fairchild was the largest individual shareholder of IBM and owned several companies that were in business with the military. By the third week of September, Moore and the others were ready to sign an eight-year contract establishing the Fairchild Semiconductor Corporation.§ The dissidents left Shockley and became known as 'the traitorous eight.'¶

Moore's timing was perfect. The transistor race kicked into high gear on October 4, 1957, when the Soviet Union surprised the world by launching the first artificial satellite.** The United States tried to replicate the achievement on December 6 with its Vanguard Test Vehicle, which after exploding during launch was nicknamed the 'flopnick.' The US military needed transistors for the space race and was willing to pay $100 or more for a transistor meeting the

---

* Together with his wife Rose.
† A transistor created by diffusing impurities into the material using gases at high temperature.
‡ Coyle and Rock had already done a lucrative transistor company deal.
§ Which was going to be part of Fairchild Camera, a division that was already selling to the military.
¶ This included the seven original dissidents plus Robert Noyce, who Arthur Rock insisted should come because of his leadership skills.
** In 1957 the Soviet Union produced 2.7 million transistors.[42]

right specifications.* Fairchild Semiconductors knew this, but so did everyone else.

In the three years since Shockley met Moore, the number of transistors manufactured in the United States jumped from 1.3 to 28.7 million. But that was still a drop in the ocean compared to the more than 456 million vacuum tubes produced by the United States in 1957. At Fairchild, Moore and his team were determined to become the first company to manufacture a diffused silicon transistor. They eventually reached success with the 'mesa,' a transistor manufactured on silicon wafers sliced out of a cylindrically grown crystal. Mesa transistors were manufactured in batches, but they had to be cut and put in a casing. That slowed down the manufacturing process.

Soon the team at Fairchild realized that cutting and connecting electronic components was a waste of time. Jean Hoerni, a Swiss-born physicist and member of the traitorous eight, developed a planar manufacturing process that allowed printing multiple components in the same wafer. This paved the way to integrated circuits.†
By the end of the decade, the bet of the traitorous eight had paid off. Fairchild went from manufacturing 100 transistors a month in the lab to about 100,000 a week on the factory floor. Moore's law was off to the moon.

*

---

* Bob Noyce went after that market. The charismatic physicist with a PhD from MIT made a deal to provide parts for a supersonic nuclear bomber: the B-70 Valkyrie. Their new company had no credibility in that space, but since the contract went through IBM they could leverage their link to Sherman Fairchild to close the deal. They soon got an order for 100 transistors at $150 each (more than $1,500 in today's money).
† Bob Noyce introduced the idea at Fairchild in a memo sent in January 1959. But the first integrated circuit was not created at Fairchild but in Dallas, Texas, by Jack Kilby in 1958, initiating a decade-long patent battle with Fairchild. For his discovery, Kilby was awarded the 2000 Nobel Prize in Physics. Noyce and Hoerni were not alive to share the prize.

The early days of transistor manufacturing were quite different from the Liberty ships described by Rapping. The Liberty ships manufactured at the end of the war were almost identical to those manufactured at the beginning of it. But the transistors manufactured by Fairchild in 1959 had little in common with the transistors produced at Bell Labs a decade earlier. The transistor was not a thing, but an idea that took different shapes and forms. That means that the growth of transistors cannot be compared directly with the airframes studied by Wright, or the ships researched by Rapping. Transistor manufacturing transcended teams. It was the result of an industry where each new incarnation of the idea resulted in a new learning curve. A curve that soared higher than the one before (Figure 6).

This is what explains the difference between the experience curve of Moore and the learning curves of Thurstone. Moore's law can break Thurstone's ceiling because it is a combination of learning curves. As each learning curve peters out, it is replaced by another curve that soars even higher. A new 'generation' of the technology. This combination of learning curves is what gives rise to Moore's famous exponential* growth.

This process was prevalent during the early days of the transistor. In fact, Shockley did not build the first transistor. That honor goes to his lab members Bardeen and Brattain, who in 1947 developed the original point-contact transistor. That design was replaced within weeks by Shockley's bipolar junction, which he created in a frenzy during the end-of-the-year holidays to get back into the race. These first germanium transistors, were then replaced by silicon transistors, which themselves evolved from Fairchild's mesas to Hoerni and Noyce's integrated circuits. The industry moved faster than any team, not along a single learning curve, but along competing curves. People in Silicon Valley, Paris, and Moscow were all

---

* An exponential function is one in which the quantity in question grows multiplicatively instead of additively (which would be the case of a linear function). That is, instead of growing 10, 20, 30, 40, 50, etc., it grows 10, 100, 1,000, 10,000, 100,000, etc.

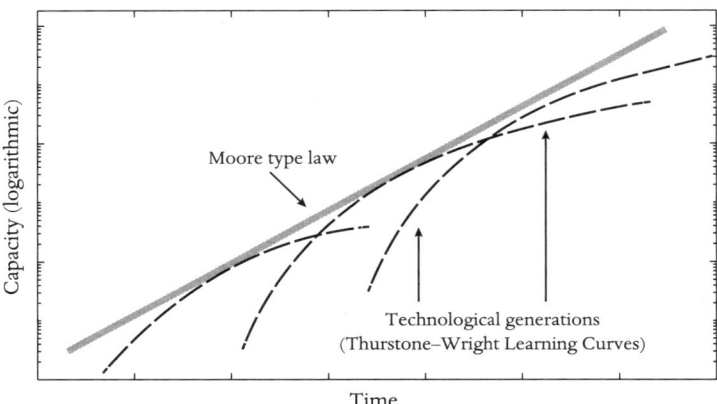

*Figure 6: Moore-type law emerges as a combination of multiple generations of Thurstone-type learning curves.*

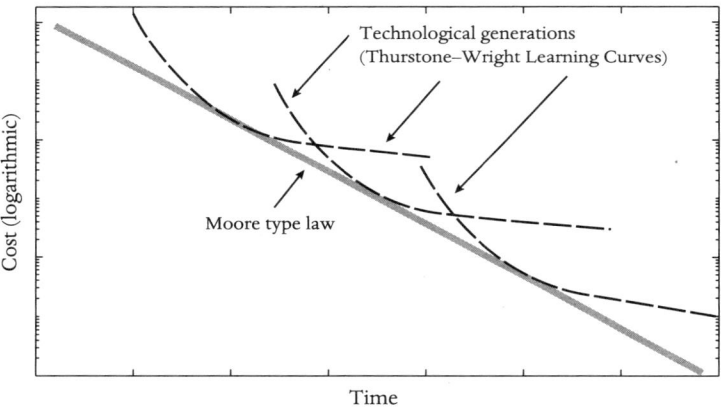

*Figure 7: Moore-type law emerges as a combination of multiple generations of Thurstone-type learning curves.*

taking stabs at the same problems. And where some failed, others succeeded. Together, these curves combined into the exponential growth that Moore popularized in 1965. Thurstone, Wright, and Rapping were right. But Moore was looking at something different.

In 1965, Moore was a veteran of this relay race. But so were many others. Just as the traitorous eight left Shockley in 1957, a team of twelve left Fairchild to found Rheem in 1959.* Hoerni also left Fairchild in 1961 with two more of the traitorous eight to form Amelco (today's Teledyne). Seven years later, Moore, Noyce, and Andrew Grove, a Hungarian chemical engineer, would leave Fairchild to create Integrated Electronics, or Intel.† The following year, Jerry Sanders, along with seven colleagues, would leave Fairchild to form Advanced Micro Devices or AMD. Ironically, Shockley's biggest mistake might have been his most valuable contribution. By threatening to leave Beckman in front of his team he validated a culture of spinoffs that is still prevalent in Silicon Valley.

Transistors, however, are just one of many technologies following a 'Moore-type' curve. Thirty-one years after Moore published his eponymous law, William Nordhaus, a professor of economics at Yale – who would later share a Nobel Prize with Paul Romer – published what is now a collection of seminal studies on the cost of artificial light.[43,44] Nordhaus was interested in inflation estimates, which require measuring the historical relative price of products. These comparisons are challenging because products change in quality and because the utility of a product comes from the 'service' it provides.

Consider a person's need for transportation. Over time, cars have become faster and more comfortable. But they are still being replaced far some people by newer forms of transportation such as electric bikes. Replacing a twenty-minute commute in bumper-to-bumper traffic with a fifteen-minute bicycle ride involves less consumption (e.g. less expenditures in gas and car insurance). Yet, it could still represent an increase in the satisfaction or 'utility' of the transportation service.

Nordhaus focused on light because it could be used to link a consumable service across different technologies. The idea is that people do not consume candles or light bulbs, but the 'light services'

---

* Their timing was terrible: they left with mesa cookbooks a few weeks before Hoerni succeeded with his planar design.
† Again with the financial support of Arthur Rock.

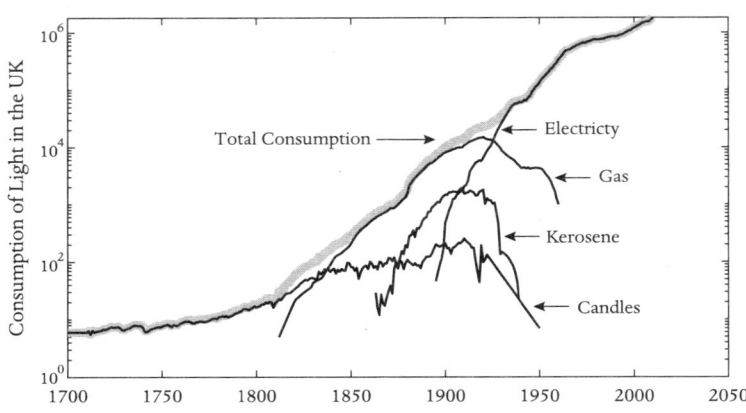

*Figure 8: Consumption of light in the United Kingdom. From Fouquet and Pearson.*[45]

they provide. These light services can be used to equate different illumination technologies.

Nordhaus found 'improvements in lighting efficiency to be nothing short of phenomenal.'[44] Paleolithic oil lamps provided a ten-fold improvement over open fires. Between 1800 and 1992, lighting efficiency improved by a factor of 900. It followed a Moore-type law, not as fast as that of transistors, but one that had been going for centuries. As in the case of the transistors, these improvements came from improvements in technology contributed by different teams.

Artificial light follows a Moore-type law that doubles every fourteen to fifteen years. This fact has been well documented by economic historians specialized in the history of energy, such as Roger Fouquet and Peter Pearson.[45] They've shown, for instance, that between the years 1700 and 2000 the consumption of light in the United Kingdom went from 10 billion lumen-hours to over a million billion lumen-hours. To explain the magnitude of this change, consider this tidbit that Fouquet likes to share in his presentations.* In the 1400s, the cost of the light service provided by running a modern 100-watt light bulb for 100 hours would have been about

---

* He was recently an invited speaker at my research group's seminar.

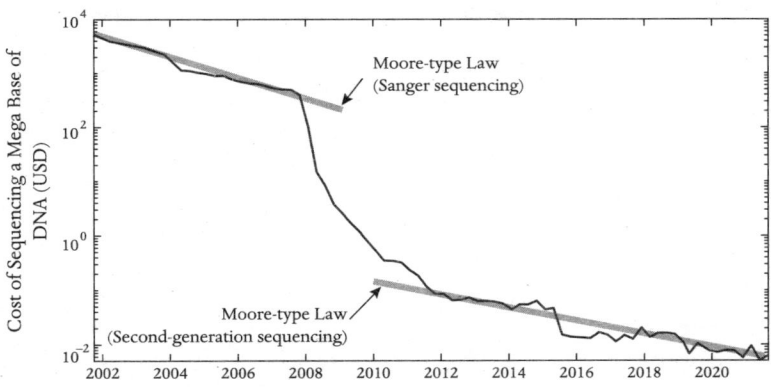

*Figure 9: Cost of sequencing 1 million bases of DNA in USD.[47]*

£10,000 in today's money. Experience curves might be the closest we will ever get to alchemy.

The evolution of solar panels follows a similar story. They also began at Bell Labs,* leveraging *p-n* junctions similar to the ones used to create transistors. On April 25, 1954, Bell Labs announced solar panels at an event where they powered a toy Ferris wheel and a radio transmitter. Since then, photovoltaic cells have continued to become cheaper and better as a consequence of continuous innovations, dropping from $105 per watt in 1975 to about ¢20 per watt in 2020.[46]

But what about artificial intelligence? Once again, the intelligence we are generating in silico mimics the patterns we observe in reality. Single models, such as GPT-2 or GPT-3 eventually reach a limit, just as Thurstone predicted. But these limited models are quickly replaced by new ones, like GPT-4 or o1, models based on more data and new ideas that reach new heights. And while they are also limited, they combine with the previous models to produce a Moore-type law.† The growth of AI also involves multiple technological generations. Generations that sometimes last less than a year.

---

* With the work of Russell Ohl, a semiconductor researcher, and Daryl Chapin, Calvin Fuller, and Gerald Pearson.
† Some people talk about a Hyper Moore's Law, but this seems to be more of a marketing trick. For the purpose of this book, any growth that follows roughly

Interestingly, technology can sometimes outpace Moore's predictions. During the early days of genetic sequencing, the cost of reading 1 million letters of DNA was around $5,000. In 2021, reading 1 million letters of DNA costs only half a cent![47] Gene-sequencing technology moved from one experience curve to the next, as sequencing centers transitioned to 'second-generation' technologies in the late 2000s.

In the world of Thurstone and Rapping, Moore's law is not trivial. But the world is not a collection of ship-making teams. Individuals, teams, and firms peter out. But at large enough scales, we can break Thurstone's ceiling. That's why the cost of computer chips, solar panels, and gene-sequencing technologies can continue to fall. Sustaining experience curves such as Moore's law, however, is not easy. In fact, there is good evidence showing that these efforts can become progressively harder. Recent studies have shown that *'the number of researchers required today to achieve [Moore's law] famous doubling . . . is more than 18 times larger than the number required in the early 1970s.'*[48] The teams needed to support the growth of Moore-type laws have grown increasingly larger in a world where, as some economists argue, ideas may be getting harder to find.[48] These large teams, however, continue finding room to grow in that deep quantum *bottom*,* with some firms now getting closer to the Armstrong[†] scale.[50]

---

an exponential is considered a Moore-type curve, no matter if it is an exponential that doubles every year or every day. Growth that follows roughly a power, is considered a Thurstone–Wright–Rapping-type curve.

* This is a reference to Richard Feynman's famous 1959 lecture, 'There's Plenty of Room at the Bottom'.[49] In his signature style, Feynman explained that electronics had a long way to go because components could be made much smaller, one day reaching the quantum scale. This leads to the – in principle, unintuitive – idea that there is plenty of room at smaller scales or at the bottom (instead of the intuitive idea that there is room at the top or at larger scales).

† An Armstrong is a unit of measurement equal to $10^{-10}$ m, which is associated with the scale of an atom.

So, the passing of the proverbial baton continues. And, for firms and teams, this relay race can be ruthless. It is the story of David repeatedly defeating Goliath. The story of transistors displacing vacuum tubes, cars replacing horse buggies, and digital cameras replacing chemical photography. A strange but not uncommon dynamic where big corporations go out of business just as Hemingway predicted: first gradually and then suddenly. *

Early in the game, Shockley was king. So probably were tallow candle manufacturers. At first, newcomers are nowhere near as good as incumbents. But when the learning curves cross, and the levee breaks, it is often too late for a king to reclaim his throne. But why cannot incumbents adapt to that sudden change of fortune? Why does history continue to repeat itself? To understand this, we need to explore the history of disruptive innovation.

* Like Ernest Hemingway described in *The Sun Also Rises*.

# 4.

# Gone with the Web

Every professor has a few stories they use repeatedly to illustrate their biggest ideas. In the case of Clayton Christensen, one of those stories was that of steel.[51]

Clay started many of his talks by noting that during most of the twentieth century there were two ways to produce steel. Early on, different types of steel were produced in massive factories that benefited from economies of scale. Some companies, however, used mini mills. These were comparatively tiny furnaces that could produce steel at 80 percent of the cost of larger operations. The problem with mini mills was that in the beginning they could only produce low-quality grades of steel.

Steel is not all the same. It comes roughly in four tiers. At the bottom you find the lowest-quality forms. These are the bars used to reinforce concrete, or rebars. Rebar is easy to make, leading to dog-eat-dog levels of competition. At the top tier, you find the steel sheets used to manufacture cars and appliances. These are much harder to make but are considerably more lucrative.

When mini mills were introduced, they entered the bottom of the market. Large manufacturers were happy to exit the rebar industry, increasing their profitability by focusing on higher-tier products. But as mini mills continued to climb the quality ladder, they began making more advanced forms of steel. First it was rebar, then bars and rods. Eventually, mini mills began making structural steel. By the time they got to steel sheets it was too late for large producers to adopt mini-mill technology. Like the proverbial boiled frog, large steel producers sat quietly as mini mills took over the market.

Christensen's theory of innovation is filled with examples of ragtag entrants repeatedly defeating powerful incumbents. The script is almost always the same. Toyota did not enter the United States with luxury vehicles, but with the Corona, a small subcompact designed for price-sensitive consumers. When Steve Jobs and Steve Wozniak began making computers, they did not try to compete with IBM's massive mainframes. The Apple I was an 8-bit computer for hobbyists built inside a briefcase.

New technologies repeatedly bring established companies to their knees, not because these technologies are better to start with, but because they are worse. Sony's pocket transistor radio (the TR-63) had an appalling sound quality compared to the furniture sized vacuum-tube radios that sat proudly in American living rooms in the 1950s. But technologies that start out worse can get better. Like the mini mills, Sony's transistor radios eventually became good enough to enter the high-quality radio market dominated by RCA.

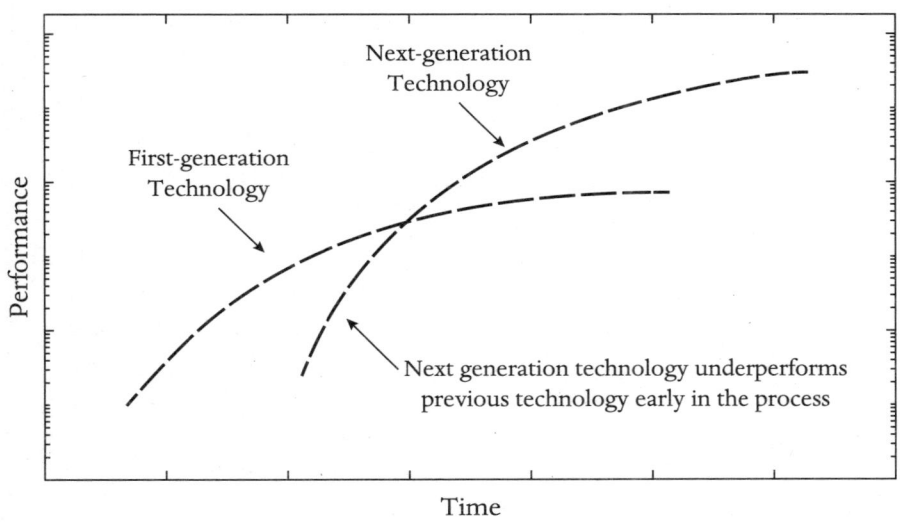

*Figure 10: How overlapping learning curves explain disruptive innovation.*

The point of Christensen's story is that we can think of disruptive innovation – situations in which a seemingly inferior technology takes over the market – in terms of overlapping learning curves (Figure 10). A 'superior' technology, that has already achieved its peak performance – such as vacuum-tube radios – is suddenly replaced by an inferior variant with higher potential (e.g. transistor radios). What makes this counterintuitive is that the initial under-performance of newer technologies is what allows them to take incumbents by surprise. But is this the full story? Or is there something else going on? After all, shouldn't an executive knowledgeable about Christensen's theory be able to anticipate this sequence of events? Or is there something limiting their ability to adapt?

Let's look at a more recent version of this story.

In September of 2000, Blockbuster – at that time the leading video-rental company in the United States – declined an offer to buy Netflix* – an up-and-coming website dedicated to selling and renting DVDs. Netflix's founders, Marc Randolph and Reed Hastings, had been desperately waiting to meet Blockbuster's executives. The fledgling startup was running out of cash, since they had not found a way to get enough people to rent their DVDs.

Marc and Reed were at their company retreat when they received a call from Blockbuster's general counsel. Blockbuster's executives wanted to meet them in Dallas the next morning. The call came at an awkward time. The annual retreat was an occasion that, as Marc recalls in his book *That Will Never Work*, Reed drank and made it count. During their retreat, the team was indulging in what had become a company tradition: employees dressed up as characters reenacting famous movie scenes. Reed was dressed as Kirsten Dunst in *Bring It On*, a movie about a cheerleader competition that had been one of the biggest releases of that year.

Marc, Reed, and Netflix's CFO, Barry McCarthy, began immediately

---

* The Netflix content from this chapter comes from Marc Randolph's memoir: *That Will Never Work: The Birth of Netflix and the Amazing Life of an Idea* (Hachette UK, 2019).

looking for flights. They were unsure how they were going to make it to Dallas. Reed suggested flying private.

'That's gotta be at least a twenty thousand round-trip!' Barry cried.

Reed countered by saying they had 'waited months to get this meeting and were on track to lose at least $50 million that year.'

So, at the crack of dawn, they rolled into Santa Barbara airport where a Learjet 35A owned by Vanna White – one of the historical co-hosts of *Wheel of Fortune* – was ready to take off.

The team was not dressed for a meeting of that caliber. The Texan AC pierced through Barry's Hawaiian shirt. The Blockbuster team was led by John Antioco, a CEO who had become famous for turning around struggling companies such as Taco Bell and Pearle Vision. He had a knack for identifying the core aspects of a business and restoring a company's morale. In the year 2000, Antioco was brimming with confidence. He had taken Blockbuster public a year earlier, selling 18 percent of the company for more than $450 million.

Reed led the pitch. He served Antioco's team a classic compliment sandwich. This is a traditional American three-part argument where the most difficult part of the message is hidden between two compliments.

He started by complimenting Blockbuster's employees and vast network of locations, then proceeded to indicate that Blockbuster could benefit from 'the expertise and market position that Netflix had obtained' on the web. He concluded by inviting the two companies to join forces by letting Netflix run the online part of the combined business.

Antioco pushed back. He argued that the dot-com hysteria was exaggerated and that the business models of online ventures were unsustainable. Eventually, he asked them for a price. 'If I were to buy . . . what are you thinking? For a number?' Breaking the awkward silence, Reed suggested 'Fifty million.'

Marc recalls seeing something new in Antioco's face. An involuntary tweak on the corner of his mouth. Later he would conclude that Antioco was trying not to laugh.

But I don't need to tell you how the story played out. Today, Netflix is a global company valued at more than $300 billion. Blockbuster went from having over 9,000 stores in 2004 to less than 100 a decade later.* At some point, Blockbuster did try to rent DVDs through a website, but they failed. So why was the multibillion-dollar company unable to compete with Marc and Reed's startup?

Christensen's theory might not be the best way to explain this. Netflix did not enter the market by providing lower-quality movies or by targeting less sophisticated or more-price-sensitive movie lovers. In principle, both Netflix and Blockbuster had the same goal: putting a good movie in the hands of a customer. But in practice, they were satisfying that need through very different means. While online and offline business can seem similar, they require drastic changes in organizational structure, changes affecting what some scholars call 'architectural knowledge.'

In 1990, two colleagues of Clay Christensen, Rebecca Henderson and Kim Clark, published a study formalizing the idea of architectural knowledge.[52] This is not knowledge about items, such as DVDs, or knowledge that is carried by individuals, such as store managers, but knowledge embedded in an organization's network of interactions. That makes architectural knowledge highly tacit, and hard to see, manage, or copy.

Blockbuster and Netflix were both in the business of renting DVDs, but Blockbuster's knowledge was on how to manage a franchise involving thousands of physical stores. Its store clerks knew how to run a local business but had no experience of shipping directly to consumers. That requires a change in the pattern of interactions among employees that is hard to know in advance and almost impossible to engineer. Netflix, on the other hand, had been discovering that pattern of interactions from the very beginning. They had put enormous amounts of thought into things that Blockbuster employees never had to worry about, from the design of the

---

* See www.hollywoodreporter.com/news/general-news/blockbuster-close-remaining-300-stores-653802/

shipping envelopes to the best way to manage a floating inventory of airborne DVDs. Netflix employees never had to deal with a customer face-to-face, but they knew how to put DVDs in envelopes like kooky elves at Santa's workshop. Blockbuster lacked any of that experience. Developing an online DVD rental service required a rewiring of their organization that went beyond their managerial capacities.

A similar story took place a few years earlier when Jeff Bezos – the legendary founder and CEO of Amazon – met the executives of his sector's giant: Barnes & Noble.[53] After reading a *Wall Street Journal* article, Len and Stephen Riggio invited Bezos for a steak dinner at Seattle's Dahlia Lounge. The Riggios told Bezos that they were going to launch a website that would crush Amazon, but that since they liked what Amazon was doing, they were open to possible collaborations.

Over the phone, Bezos told the Riggios the collaboration was unlikely to work. He was a big believer in the ability of small companies to triumph, and understood that, while similar on paper, Amazon and Barnes & Noble were immensely different under the hood. Unlike Barnes & Noble, whose inventory was spread all over the country, Amazon held its inventory in a single location, without the need to display it nicely to please their customers. Amazon also had much better economics. Their fixed costs were tiny compared to those of brick-and-mortar stores, allowing them to aggressively reinvest their revenue. Eventually Barnes & Noble did open an online store, going into a full-blown war against the startup. Days before Amazon's IPO, Barnes & Noble filled a lawsuit against Amazon in federal court contesting the idea that Amazon was the world's largest bookstore. During the following months, the two companies competed in selection and prices. But Bezos was confident that the Riggios would face serious challenges competing online.

The logistics of a direct-to-consumer business are very different from those needed to run an army of physical stores. Barnes & Noble knew how to send large shipments to physical stores using distribution centers that packed trucks with boxes containing multiple copies of the same books. Shifting from that to shipping a single book to an individual anywhere in the United States was not easy. For Barnes & Noble the web was a secondary channel, which made the Riggios reluctant to put their best people on what could be a money-losing operation. Both Barnes & Noble and Amazon had plenty of books and customers. But transforming Barnes & Noble's luxurious brick-and-mortar stores and wholesale-oriented distribution centers into the type of operation that Amazon was running from a derelict Seattle building required more than building a website. It required a complete rewiring of the organization.

In their study, Henderson and Clark use Xerox as an example. Like Blockbuster and Barnes & Noble, Xerox was once the dominant firm in its industry: photocopying machines and toner. But its business model, based on large machines, began to suffer when competitors introduced smaller copiers. This is compatible with Christensen's theory of disruptive innovation, but it is also in line with Henderson and Clark's lessons on architectural knowledge.

To understand why shifts in architectural knowledge are difficult, Henderson and Clark recommend looking into two key concepts. One is the idea of a canonical design. The other one is a theory of the structures used by an organization to accumulate knowledge.

A canonical design is defined by the way in which different components are organized in a product. Think of a car. A top-of-the-line Mercedes and a beaten-up Corona share many features. They both have four wheels, two-to-four doors, and a steering wheel in about the same place. Innovations that involve improvements on one of the components, let's say a new set of wheels, are not architectural if they don't require changing the relationship among other components. Architectural innovations require such changes. Clark and Henderson use airplanes as an example. In the early days of flight, changing one combustion engine for another was a non-architectural innovation that most companies were able to adapt to. But the replacement of propellers with jet engines was different. It required a redesign of the entire airframe. This was something that not all companies were able to adapt to. An exception was Boeing, which in 1957 became an industry leader by introducing one of the first successful passenger jetliners: the 707.

What Netflix and Amazon had that Blockbuster and Barnes & Noble didn't wasn't a website. What they had was architectural knowledge on how to operate on a direct-to-consumer model focused on fulfilling individual orders directly from a single distribution center. Knowledge that went beyond the product and was about the process.

But why is adapting to architectural change so difficult?

Remember the network model of organizational learning advanced by Linda Argote? The one where learning involves rewiring networks connecting people, tools, and ideas. Argote's networks, or 'knowledge reservoirs,' provide a great way to visualize the architectural changes described by Henderson and Clark. We can think of the distance between two organizations, such as Netflix and Blockbuster, as the number of links we have to rewire to make their networks similar. For incumbents, such as Blockbuster, the introduction of an innovation, such as online DVD distribution, may look simple, but it is a reorganizational nightmare for a workforce that is distributed across thousands of physical stores. To approximate Netflix's network, Blockbuster would need to rewire its network of people, tools, and tasks. That distance, in Argote's theory, was larger than what Antioco might have imagined.*

Now, before we conclude this chapter, it is important to note that not 'everyone is either disrupting or being disrupted,' as was pointed out in a poignant essay by the Harvard historian Jill Lepore.[54]

Lepore's essay criticized Christensen's work by carefully picking apart many of his examples. She begins with the history of hard drives, the subject of Christensen's 1992 doctoral thesis. Lepore

---

* Of course, not all these networks are equally difficult to change. Some of these networks, like those involving tools, can adapt more easily. Connections between tools and tasks can be translated into protocols and procedures. They are less tacit and more explicit. The networks involving people, however, are often tacit, making them somehow inflexible to important changes in context.

notes that the first hard drive 'was invented at IBM, in 1955, by a team that included Alan Shugart,' a man who then founded a company that in 1976 'introduced a 5.25-inch floppy-disk drive,' and in 1978 introduced an 'an eight-inch hard-disk drive.' After selling that business to Xerox, Shugart founded the company that we now today as Seagate, introducing 'the first 5.25-inch hard-disk drive' in 1980. Lepore takes issue with the fact that Christensen classified Shugart Associates and Seagate Technologies as new entrants into the disk-drive business, since these were companies founded by people who had clear experience in the technology.

Lepore's essay then digs into Christensen's work on the 1950s mechanical excavator industry. Again, she notices that disruptors were incumbents who had been making cable-operated shovels from the beginning of the twentieth century. Finally, she goes after Christensen's mini-mill story, arguing that a key difference between large and small steel producers in the 1980s was that large steel producers had unionized labor forces, which were mostly absent in smaller mills.

Lepore is right in pointing out Christensen's historical inaccuracies. But these examples do not deny the existence of disruptive innovation. Christensen's sin was to overemphasize the idea that disruptors were outsiders, when, upon closer examination, they were either incumbents or firms with 'related knowledge.' We will dedicate an entire chapter to the idea of related knowledge later in this book. But for now, it is worth noting that both Lepore and Christensen have a point.

Lepore was right to point out that these industries were not being disrupted by complete outsiders, but by firms and people that were clearly in the orbit of the key technologies, much like when Apple introduced the iPhone. But Christensen was also right to identify disruptive innovation as a key phenomenon, even if this is technically a battle among a few well-placed incumbents.* Disruption

---

* In fact, even Lepore partly fits the bill. Even though she is not a complete disruptor of Christensen's work (since she does not provide an alternative mechanistic

may not come from as far as Christensen might have argued, but as a general idea, it still has legs.

Before publishing *The Innovator's Dilemma,* Clay Christensen received a call from the Hungarian-born chemical engineer and Intel co-founder, Andrew Grove. Grove had become the company's chairman, and invited Christensen to explain his research after having read one of his papers.[55]

But by the time Christensen arrived, Grove's plans had changed. 'We have only ten minutes for you. Tell us what your model of disruption means for Intel.' Christensen declined to cut his presentation short, and eventually convinced Grove to let him start with his mini-mill story. At the end of the presentation Grove went on to articulate their strategy for the bottom of the market. It focused on Celeron processors, a series of low-cost CPUs that Intel sold between 1998 and 2023.

Christensen was often reluctant to tell people what to do. He liked presenting his models using examples from a different industry to trigger an *aha!* moment in his listeners' minds. He knew that once people got the model, they would articulate a better answer to their problem than he could. His desire was to teach people not what to do, but how to think. Funnily enough, conceptual knowledge is often hard to transmit by talking about pure concepts. It is a form of knowledge that travels better in a story.*

———

model to replace that of Christensen), she is close to Christensen in many key dimensions. She is a professor at the same university where Christensen worked, and earlier in her career worked at Harvard Business School under Michael Porter.
* There is good evidence that stories are recalled more easily and affect a person's beliefs more effectively than facts, making them a powerful form of communication.[56]

5.

# *Forget Me Not*

In 2017, the British newspaper the *Guardian* published an article on a curious trend: the price of Elvis memorabilia was plummeting.[57] Collectors who had spent their life hoarding records and souvenirs were now facing a shrinking market. The usual buyers for these items, other collectors, were dying of old age, leaving their children with heaps of Elvis memorabilia that they were struggling to sell. Records that in the '90s fetched over $100 were being sold for one tenth of that price. The article concluded that as diehard Elvis fans joined him in heaven, the collective memory of 'The King' was beginning to fade.

So far, we have focused on the growth of knowledge. But growth is not guaranteed. Sometimes, the dynamics of knowledge is dominated by forgetting, a phenomenon that also follows empirical laws.

In 2016 I took my team of graduate students and postdocs to the Chilean city of Valparaíso. At that time, I was an associate professor at MIT. My goal was to help a Chilean university's new PhD program by mixing their students with mine. During that month, we would work out of a startup accelerator, which for a few weeks would brim with people, ideas, and interactions. Every day I would go to the desk of each student to discuss their work. Each Friday, students would give a short talk about their research. At the end of the month, all students had to present their work during a gala event in Santiago. The four-week program was designed to push the students to hone their research skills, by forcing them to bring their work to a point where they could articulate its value to a general audience.

One of the Chilean students was Cristian Candia, a physicist working on his PhD at Universidad del Desarrollo. Later, Cristian joined my research group for a couple of years as a visiting PhD student.

Cristian was interested in music and had been playing around with data from Spotify. One day, he showed me a chart describing the popularity on Spotify of songs that had once made it to the Billboard Hot 100 ranking. As expected, popularity decayed over time, with songs from the '50s being, on average, less popular on Spotify than songs from the '70s or the '90s. But what was interesting about this chart was that the decay was not smooth. The trend had a 'knee,' an exciting bend, that couldn't be explained by a constant rate of decay.

That knee hinted to a mechanism. Maybe short- and long-term attention were governed by different dynamics. But there was a second reason why this knee was interesting. Scholars have long studied the attention received by cultural products, such as movies, patents, or scientific papers. The dynamics that are usually reported

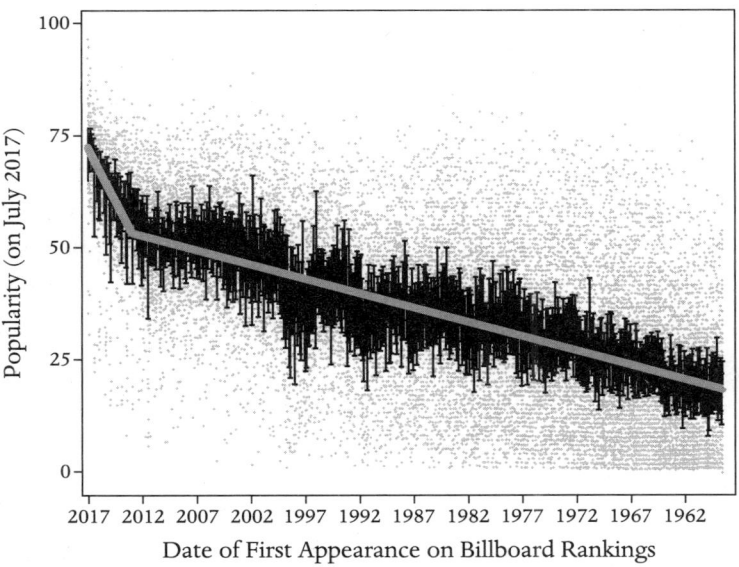

*Figure 11: Popularity of songs in 2017 as a function of the year in which they first appeared in the Billboard rankings.*

is that of a rise followed by a decay. That rise comes from the fact that attention begets attention. Even today, Elvis is more popular than many modern artists, an idea known by academics as cumulative advantage, preferential attachment, or the Matthew effect.[58-60] But our data was showing something different: a two-tiered decay that came from looking at a very biased set of songs. The sample that Cristian had selected was not representative of all songs ever produced. It was an elite sample comparing the likes of Elvis, Taylor Swift, and Elton John.[61] By focusing on songs that had reached the Billboard Hot 100 we were effectively removing that cumulative advantage. We were looking at 'comparatively' famous songs, isolating the forgetting component of collective memory.

Over the next couple of years, we explored the universality of this finding. We joined forces with my former PhD advisor, László Barabási, who gave us access to data on the citations received by thousands of papers and patents. We leveraged data on famous biographies from our online observatory of collective memory, a website that goes by the name of pantheon.world. In all these datasets we found traces of the same pattern. Attention decayed quickly, and then slowly. But why?

To explain this pattern, we built a mathematical model inspired by work on the humanities that talked about 'communicative' and 'cultural' memory. Communicative memory is sustained by people talking about a famous song, movie, or paper. The chatter that follows a famous soccer game, political election, or boxing match. A fast-decaying social reverberation. That conversation, however, feeds the growth of cultural content, such as books, videos, and magazine articles. The knee of these curves was the point in which cultural memory took over. The point at which what was written became more important than what was said.

In 2019, we published this in a paper that began with a phrase I borrowed from the Chilean poet Pablo Neruda. Every night for that month in Valparaíso, I climbed a hill* using stairs graffitied with

---

* Subida Ecuador.

one of Neruda's famous sentences: *Es tan corto el amor, y tan largo el olvido* 'love is so short, and forgetting so long'. Eventually, the stairs convinced me that this was the right way to tell that story.

Forgetting songs and movies can be an unintuitive experience. It is hard for me as a Gen-Xer to imagine people not knowing about John Lennon or Pink Floyd. But I remember a student walking into my office while I was listening to John Lennon's 'Imagine.' She thought that this was a song by Coldplay. Our society has a limited ability to accumulate knowledge, and as we move forward, knowledge falls through the cracks. Knowledge follows an 'if you don't use it, you lose it' type of principle. If you don't believe me, ask the son or daughter of an Elvis collector.

But knowledge decay extends beyond the attention received by cultural products in pop culture. It also affects the procedural knowledge that is accumulated by firms, like the shipbuilders studied by Rapping or the airframe manufacturers described by Wright.

In 1990, Linda Argote, Sara Beckman, and Dennis Epple revisited Rapping's data with a focus on forgetting. They used a model that 'discounted knowledge,' penalizing the contribution to experience of ships manufactured earlier compared to those completed more recently. They concluded that knowledge decayed quickly, at a rate of about 15 percent to 25 percent a month.[62] These estimates were later improved by Peter Thompson, who adjusted them by controlling for other factors, concluding that knowledge depreciated at a slower monthly rate of 3–6 percent. Still, if we consider this is an estimate of the knowledge lost in just four weeks, a 3–6 percent loss of knowledge is dramatic. It implies that a shipyard that stopped producing ships would lose half of its knowledge in about a year.

But the shipyards did not lose that knowledge as long as they continued to produce new ships. In fact, their production rates were large enough that learning outpaced forgetting. This means a net growth of knowledge. But in other examples, forgetting can outpace learning. That was the case of the L-1011, an advanced three-engine aircraft produced in the 1970s by Lockheed Martin.

The L-1011 TriStar was designed to be the plane of the future.

It included numerous innovations, such as glare-resistant windows, full-sized closets, wide aisles, and large overhead bins. Its unique engine configuration reduced sound in the cabin.[63] But what made the TriStar special was its advanced fly-by-wire* autopilot system. The TriStar could fly and land on its own. On May 25, 1972, it completed the first autonomous cross-country flight in the United States, bringing a crew of 115 employees and reporters from Palmdale, California, to Washington, D.C. This feature gave the TriStar a special clearance by the United States Federal Aviation Administration to land during severe weather conditions.

But the production of the L-1011 was not as smooth as its in-air performance. In fact, the story of the L-1011 has become a textbook example of organizational forgetting. The costs of producing each aircraft increased between 1975 and 1982, from less than $20 million to more than $29 million.[34] This has been attributed to inconsistent production and employee turnover. Between 1972 and 1974, the number of planes produced each year grew from seventeen to forty-one. But then dropped to a minimum of only six planes in 1977. In a world where knowledge decays, these bottlenecks can lead to periods where forgetting can outpace learning.

In some cases, the loss of knowledge can seem imminent. That is the recent history of Polaroid, a company that in the 20th century was the undisputed leader in instant photography but was steamrollered by the rise of digital cameras in the 1990s.

Polaroid was born in the roaring '20s. It was the brainchild of Edwin Land, a legendary American inventor and entrepreneur.[64] Land grew up in a modest household without too many books. One of them was a 1911 edition of *Physical Optics* by Robert Wood. Land become obsessed with this book, and particularly with the chapter on light polarization, a property that can be used to control the transparency of materials.[65]

An easy way to think about light polarization is to imagine throwing a Frisbee through a picket fence. A frisbee can fly horizontally,

---

* Electronic.

vertically, or it can wobble. But, like fitting a coin in a slot, the frisbee will only fly through the fence if it is aligned with the grating. Light is similar. Instead of a flying frisbee, it is a self-perpetuating braid of electric and magnetic fields. But like a frisbee, it can vibrate horizontally, vertically, or wobble in a circle. Polarizers are like picket fences that filter light depending on the direction of its vibration.

After graduating from high school Land was admitted into Harvard's physics program, where he continued to study optics, before quitting after completing his freshman year to create his own lab in New York. Back then, people knew how to grow polarizing crystals that were too small for real-world applications. But that changed when in 1928, at age nineteen, Land had a key breakthrough. He figured out how to build a polarizer by filling a vial of fluid with tiny crystals that he aligned using a magnet. This led him to a method by which he could produce a polarizing film by coating a thin plastic sheet with a wet layer of microscopic crystals that he then aligned by stretching the drying flim. This led to Land's first patent, which he filed in April, 1929.

With a patent under his arm, Land returned to Harvard. He began collaborating with George Wheelwright, a recent Harvard graduate from an elite New England family. But, again, Land would not stay at Harvard for long. In 1932, he became an official dropout and a co-founder of the Land-Wheelwright Laboratories, a company focused on commercializing his polarizing film. In July of 1933 Land's patent was approved and the business took off.

The startup closed contracts with Bell Labs and Eastman Kodak as Land learned to sell his inventions with flare. In a legendary sales meeting, he invited American Optical executives to a hotel room in Boston where he had placed a fishbowl on the windowsill. When the executives entered the room at the agreed upon time, they were blinded by the light refracted by the fishbowl. Land immediately handed them his polarized filters.

In 1937 Land-Wheelwright went public as Polaroid, putting Land's creative genius on high gear. Polaroid introduced windows with tunable darkening that worked by rotating two sheets of polarized glass. At the 1939–1940 New York World's Fair, they introduced 3D movies by using polarized filters to beam different portions of a film to each eye. This 3D technology soon found a wartime application, by producing air photography that – when viewed with 3D glasses – showed relief. But the war also pushed the ingenuity of Polaroid's chemistry division. When Japan invaded the island of Java, limiting the supply of quinine, a key ingredient in polarizers, two Polaroid chemists figured out how to synthetize it. The team included Robert Woodward, who went on to synthetize multiple organic molecules such as cortisone and cholesterol, and won the 1965 Nobel Prize in Chemistry.

During the war, Polaroid's business exploded, from $761,000 a year to more than $16 million, with most of that income coming from military contracts. But as the war ebbed, Land began to worry about his 1,200 employees. The legend says that during a vacation in Santa Fe, New Mexico, Land took a picture of his three-year-old daughter Jennifer, who asked why she couldn't see the picture right away. Back then, film had to be sent by mail to Eastman Kodak, a company built on the discovery of dry film. Kodak would develop the film and send it back in an envelope in about a week. That legendary night Land went into a creative frenzy, pacing around the resort while thinking about instant photography. It was a formidable technical challenge, packing an entire darkroom of highly reactive chemicals in an envelope that was sturdy enough to be shipped around the country. It fitted Land's motto: Don't undertake a project unless it is manifestly important and nearly impossible.

In 1944, a few months after the Santa Fe trip, Land produced an instant photograph, and by 1946, his team was making decent test pictures every week. The great reveal came on November 26s, 1948, the day after Thanksgiving, when the Polaroid's sales team took fifty-six Model 95 cameras to Jordan Marsh, a big department store in Boston, and offered free pictures to the shoppers. They sold all fifty-six cameras that day.

As is often the case, not everyone loved Land's invention. Some people ridiculed instant cameras as toys unable to produce 'real photography.' But you know how that story goes. Land was a disruptive innovator who had successfully jumped into a new learning curve. The quality of Polaroid instant film would grow to the point of making it a favorite of renowned artists such as Andy Warhol and Ansel Adams. And the technology needed to produce the film became so nuanced that Kodak would be unable to reproduce it without infringing on Polaroid's patents. In 1991, the year that Land died, a federal court ordered Kodak to pay Polaroid the largest recorded settlement in US history, a legendary $925 million.[66]

But not even that settlement could save Polaroid from what was coming. In the 1990s, Polaroid was being disrupted by digital photography, a technology decades in the making.* This time Polaroid was the one ridiculing the technology that would lead to their bankruptcy. A few years later, Polaroid would stop producing instant cameras and film. The loss of knowledge seemed imminent. Land's creative fire had been extinguished, but luckily a few hot embers remained.

In 2008, at a Polaroid factory closing event in the Netherlands, Florian Kaps, an Austrian entrepreneur, met André Bosman, a manager with three decades of experience at Polaroid's Enschede plant. Kaps was deep into instant photography. He owned an online shop in Vienna that sold instant film to thousands of customers around the world.[68] Kaps was afraid that he was going to run out of film and wanted the Enschede plant to keep going. Bosman was allegedly sent to dissuade Kaps, who might have mobilized the workers, but over a few beers the Austrian convinced him to join his crusade.

The epiphany came on a Friday afternoon, when the scrapping of key equipment was scheduled for the following Monday. Kaps and Bosman spent the weekend lining up money and contacts to keep the plant going. They used every trick in the book and eventually convinced Polaroid executives of selling the equipment and

---

* Going back as far as 1880, when Shelford Bidwell used electric signals to transmit a drawing of a butterfly captured by a selenium cell.[67]

leasing the building. The deal was made possible with the financial support of Marwan Saba, a friend and supporter of Kaps. Together, Kaps, Bosman, and Saba took over the plant with their newly incorporated firm: The Impossible Project.

The trio got to work by hiring about a dozen laid-off employees. These were not slim pickings. I asked Kaps this question directly when I had the chance to meet him on a trip to Vienna. He told me that Bosman put together a 'dream team,' manning each part of the production process with the best person the plant had. But even this dream team struggled to keep Land's legacy alive. The company was staying afloat by selling the remaining stock of Polaroid film, which they had bought together with the plant. But even though this was enough to keep them running for a few years, they needed to figure out how to make new film. The Enschede plant, however, did not have all they needed. The operation relied on suppliers for chemicals that had been discontinued or banned. So even though they had the original equipment and employees, the loss of knowledge was becoming catastrophic. They began to produce instant film of poor quality, which was hard to use and would easily fade. Users of the 'new' film had to immediately shield it from any light after removing it from the camera to avoid overexposure. Pictures also took about forty minutes to fully develop, bubbling up regularly, failing frequently, and fading rapidly.[69] Land's embers were still warm, but not hot enough to reignite the flame.

Kaps had the best of intentions. In the 2020 documentary *An Impossible Project*, he is shown as the world's sweetest contrarian.[70] And he is not a luddite either. His passion for saving vintage technology sometimes requires new methods. But his fascination with what we now see as retro did not come with anything close to Land's engineering skills. In some ways, Kaps is the opposite of Land. He presents himself as a man fascinated by the lack of speed and scale of retro technology, even though he confesses to be a devoted admirer of the American genius. He likes to emphasize how digital cannot be smelled or touched, but Land's invention was an attempt to move in the opposite direction, into a world where photography was faster, just like digital did fifty years later. The Impossible Project was a romantic attempt to save a technology that seemed outdated, but that was nevertheless incredibly complex. It is a quintessential example of forgetting, and of how difficult it is to preserve complex knowledge.

Five years in, the Impossible Project was still unable to produce quality instant film. Some of the missing knowledge returned in 2014 with the arrival of Stephen Herchen, an organic chemist with a PhD from MIT and twenty-eight years of experience at Polaroid, including working directly with Land in the 1980s. That same year, the company was rescued by a $2 million investment from Slava Smolokowski, a former Soviet-era musician who made it big in the energy business. But that check came with strings attached. That December, Slava's son Oskar, who had joined the Impossible Project as Kaps' assistant a year earlier, became the new CEO. In a circuitous way, Kaps kept Land's dream alive, but it was time for someone else to become the steward of Land's vision.

Under new management the Impossible Project began to produce better film. In 2017, the Smolokowskis acquired the brand and intellectual property of the former Polaroid Corporation, putting Humpty Dumpty back together again. More than fifteen years after Kaps began his quixotic rescue, Polaroid is finally able to produce decent film.

We all love to think that our present is more technologically advanced than our past. After all, we live in the era of artificial intelligence and space exploration. But the truth is that not all technologies move forward. Our civilization has a limited capacity to learn, and sometimes that means knowledge is lost. The machines at the Enschede plant were a poor record of Land's genius. Kaps was able to inspire people to keep going, but that inspiration soon proved insufficient. Forgetting is as true as learning, and the reason we don't see it more often is because we tend to be relatively good at creating organizations that preserve knowledge. Our educational system, from preschool to the most advanced courses at elite universities, are efforts to preserve some of the knowledge we have created. Firms are organizations that survive as long as they continue to learn. Even this book, in its own limited way, is an effort to preserve some knowledge. But forgetting is a reality that we often gloss over. Except when everything breaks.

In his 2020 book, *Blueprint*,[71] Nicholas Christakis describes some gruesome shipwrecks. These are ghastly examples of the loss of knowledge where marooned survivors become detached from society and struggle to satisfy their most basic needs.

Consider the case of the *Batavia*, a 1629 shipwreck where the crew systematically planned the murder of women and children to preserve resources. Other macabre examples include the wreck of the *Medusa* in 1816 and *Le Tigre* in 1766, both of which ended up in murder and cannibalism. On a few occasions, survivors managed to cooperate and reach some form of deliverance. The *Doddington*, in 1755, cast a group of twenty-three survivors (all crew members) onto a rocky 47-acre island. The group survived on fish, birds, and eggs for seven months. They also had some hope, as they could see the mainland at a distance. Eventually, they sailed away on a raft built by Richard Topping and Hendrick Scantz, respectively a carpenter and a blacksmith.* But the case of the *Doddington* is an exception, where a skilled crew still struggled to support anything close to the living standards they would have enjoyed in a normal situation.

But forgetting is not always a bad thing. Sometimes, it can facilitate learning. This is exactly the story of how IBM survived the rise of personal computers and succeeded at creating one on its own.[72] By putting a team of engineers to work far from its main center of operations, IBM was able to learn a technology that was related to their business, but architecturally distant from the one being produced at its headquarters.

In 1980, William (Bill) Lowe, the director of IBM's Entry Level

---

* But even this crew didn't make it much further, as they struggled to continue their journey up the coast of Africa.

Systems in Boca Raton, Florida, was worried about the rise of personal computers. He understood these could disrupt IBM's mainframe business. So, in a bold effort, he managed to convince IBM's Corporate Management Committee to support a team of twelve engineers to work on a personal computer. The key to Lowe's 'Project Chess' was that it was not going to be executed in Armonk, New York where IBM was focused on large business machines with ominous names, such as the Datamaster. Project Chess would take place in Boca Raton, Florida. Not only would it be far from IBM's headquarters, but it would use components sourced from outside vendors, leveraging knowledge that was outside IBM. That was also a bold move for a company that was used to working with internal knowledge. In a matter of months, the team of engineers was able to put together a personal computer which they showed to the company in January of 1981. We can think of this as the birthday of the PC. Sometimes forgetting can go a long way.

## The First Law: The Principles of Time

Thurstone's students, Moore's law, and Land's film, all carry valuable lessons about knowledge. We now know how learning curves combine into experience curves, and how that explains histories of disruptive innovation. We have also learned why it is hard for organizations to jump from one learning curve to the next, as the architectural knowledge they hold is hard to redeploy. This has provided us with a mechanistic view of disruption, the process by which inferior products and processes – from instant cameras to miniature steel mills – beat juggernaut incumbents. But we have also learned that the growth of knowledge is not always guaranteed. Learning curves face headwinds that can gnaw away at their knowledge quite rapidly when production slows downs or ceases completely. At first, this decay can be fast, involving the ephemeral knowledge sustained by repeated acts of communication. But if enough knowledge becomes codified, that decay enters a second regime, where knowledge continues to disappear at a slower rate.

Together, learning curves, experience curves, disruption, architectural knowledge, and forgetting, provide a simple but limited principle to understand the dynamics of collective learning. These lessons, however, involve examples that are limited to single tasks or industries. They tell us about shipyards making ships or computer algorithms learning to recognize the objects in an image. But understanding the growth of knowledge requires more than understanding how people produce more or less the same thing faster and better. We also need to explore how knowledge moves among geographies and activities and to study the importance of complementarities in the accumulation of knowledge.

This makes learning curves, experience curves, and forgetting curves an easy point of departure that deliberately avoids the complexities of the infinite alphabet. The uniqueness and diversity of knowledge plays a muted role in research focused on how to do more of the same thing. But the growth of knowledge is not just about bakers making more and better bread, or chemists doping silicon more effectively. Teams, companies, and nations often need to learn new things. Where does that knowledge come from? And how does the knowledge in their environment affect their ability to learn? How did the manufacture of Liberty ships benefit from those working on other ship models? Or from the manufacturing of aircraft? Why did the knowledge needed to manufacture transistors appear in San Francisco, Tokyo, Moscow, and Paris, but not in Santiago, Bangkok, Quito, or Yachay? What are the principles that govern, not the growth, but the movement of knowledge across both the physical space of geography and the cognitive space of activities?

In the next chapters we will explore how knowledge moves across mountains and industries, starting with the story of one of the most 'badass' entrepreneurs in American history: the six-foot-tall father of American manufacturing, Samuel Slater.

# PART II

## The Principles of Space

# 6.

## *Yet It Moves*

As the summer of 1789 was coming to an end, and without telling a soul, Samuel Slater left the English Midlands on an ominous night.[73] He had read in the newspapers that the Legislature of Pennsylvania was encouraging the introduction of improved machines for cotton manufacture. But he also knew that bringing this knowledge to the United States was a punishable act of treason. England had stringent laws against the emigration of skilled mechanics and against shipping out of the country any drawings of manufacturing machinery. So, Slater decided to bring this knowledge to the United States hidden in his brain. He had accumulated nearly six years of experience of water-powered cotton spinning after working for Jedediah Strutt at his Milford cotton mill starting at the age of fourteen.

You may think that cotton spinning – the process of turning cotton fibers into yarn – is simple. But it was an engineering breakthrough that took decades to develop. The first patents for cotton-spinning machinery were produced in the 1730s but did not lead to any successful mills. Two attempts to implement these patents in the 1740s failed: neither the donkey-powered mill in Birmingham nor the water-powered mill in Northampton succeeded as business ventures. It took two more decades for mechanical cotton spinning to start taking shape, thanks to the work of Richard Arkwright, an English inventor and entrepreneur born in Preston in 1732, a few miles northwest of Manchester.

Arkwright was born at the right place and at the right time. The spinning jenny – a method to join multiple spindles in a single machine – was invented by James Hargreaves in Blackburn, only 10 miles away. Hargreaves also introduced important improvements to the technology used for carding, the combing process used to align fibers before spinning them.

But Arkwright was not only geographically close to these ideas. He was a connoisseur of fibers. He has trained as a barber and had established himself in 1760 as an itinerant hair merchant. During this time, he discovered a chemical process to dye hair which helped him accumulate his first small fortune. But a cotton-spinning mill was an endeavor that required more than Arkwright's 'hairy' expertise. That expertise came with John Kay,* a watchmaker born in nearby Warrington who worked with Arkwright on a cotton-spinning system based on a sequence of rollers that could spin yarn to an exquisite degree of fineness.

In 1768, Arkwright and Kay moved to Nottingham, a center for the manufacturing and trade of hosiery. Nottingham was booming. It was attracting people from all over the country, doubling its population in the five decades starting in 1750. The hosiery industry demanded yarn, and Arkwright and Kay wanted to become one of the industry's key suppliers. Once in Nottingham, they received a capital investment from the Wright bankers to erect a cotton-spinning mill. But the partnership didn't last long. The following year, Arkwright would file a patent without mentioning Kay, leading to a falling out between the two and pushing Kay to work with Hargreaves – the inventor of the spinning jenny – who was moving to Nottingham after a group of hand-wheel spinners destroyed his machines.

But Arkwright's mill did not enjoy an auspicious start. For two years, the mill, powered by horses, did not make enough progress to satisfy Arkwright or the bankers. On their advice, Arkwright contacted Jedediah Strutt, a gentleman from nearby Derby who had become rich by manufacturing hosiery. Strutt was a big consumer of cotton yarn and an accomplished mechanic and inventor, so he

---

* Not to be confused with the more popular John Kay, the inventor of the flying shuttle.

could judge both the quality of the product and the process used to produce it. Strutt understood the importance of Arkwright's spinning frame, and in 1771, together with Samuel Need, entered a partnership with Arkwright. Soon, they built a much larger mill than the one that Arkwright had in Nottingham in the small town of Cromford, taking advantage of its excellent water power.

Things finally started to work. By 1773, the mill was producing cotton yarn for the hosiery industry in Derby and Nottingham. This annoyed hand-weavers back in Arkwright's native Lancashire, but there was not much they could do to compete with the new technology. Unlike hand-spun yarn, mill-spun yarn was strong enough to resist the pull and friction of a mechanized loom. They not only produced yarn faster, but better. Arkwright doubled down, pushing the technology further by filing for a second patent in 1775 focused on improvements to the carding machinery, which together with the 1769 patent formed the basis of the 'Arkwright system.'

Arkwright and Strutt soon began replicating their technology, building additional mills in Belper and Milford. But in 1781 Arkwright and Strutt had a falling out and dissolved their partnership. Arkwright retained the mill at Cromford and Strutt the ones in Belper and Milford.

Like Arkwright, Samuel Slater was also born at the right place and the right time, in Belper on June, 1768. In 1783 he became an apprentice of Jedediah Strutt at the Milford mill. Samuel was recommended to Jedediah by his father William, who passed away in 1782. His father had told Strutt that Samuel was a good writer and mathematician of decided mechanical genius. And he was right. Samuel showed so much skill that in three years he would become a favorite of Strutt's and rise to the rank of overseer. But as more mills were built, the high-flying teenager became concerned about a saturating market and began thinking about the United States, a young republic where the technology was not yet available. So, on September 1, 1789, pretending to be a farmer and without telling a soul, Samuel Slater boarded a ship to New York, where he landed sixty-six days later.

If the American dream happened to anyone faster than to Samuel Slater, please tell me about it, because I have not heard that story. What came next was an amazing feat of entrepreneurship. In fewer than six months, the young English immigrant would become a partner of one of the richest men in New England and the undisputed cotton-spinning leader in the United States.

His story began with a quick failure in New York. Within four days of making landfall, Slater got a job at a cotton factory in Vesey Street that he quit after deeming the enterprise futile due to the lack of water power and inadequate machinery. He then learned from a sloop captain that people in Pawtucket, Rhode Island, were experimenting with the use of water power to manufacture cotton yarns. So, on December 2, he sent a letter to Moses Brown, a wealthy Providence merchant, offering his services as a master cotton spinner.

People in New England had been trying to develop cotton-spinning machinery without much success. Moses Brown was one of them. He had acquired a spinning frame built under the instruction of two Scotsmen who had observed the Arkwright system but had no real experience operating the machines. In a letter of 1789, Brown revealed having several jennies and some weavers that were unable to produce cotton wraps with 'a useful degree of perfection.' Slater's letter must have felt like a godsend.

On December 10, Slater received a letter indicating that Brown & Almy – a company formed by Moses's cousin Smith Brown and his son-in-law William Almy – needed 'the assistance of a person skilled in the frame or water-spinning.' In the letter, Brown tells Slater they had a machine that was 'too imperfect to afford much encouragement.' He also mentions that if Slater can get this machine to work, he could keep 'all the profits above the interest of the money they cost and the wear and tear of them.' In return, Slater would disclose information about how to operate the machinery.

Slater arrived in Providence on January 1, 1790, and, after examining the machines, rejected the deal. Instead, he agreed to build new ones. He requested a skillful carpenter to perform the woodwork 'under bonds to neither steal the patterns nor reveal the operation.'[74] He also requested a secure room to guarantee the secrecy of the work and was granted a dollar a day for subsistence.

The construction got started right away. Slater showed so much skill and knowledge that on April 5, 1790, only five months after he made landfall in New York, he signed a contract with Brown & Almy for the construction of two additional carding machines, a drawing and roving frame, and a spinning frame with 100 spindles. Brown & Almy would provide the capital and Slater the knowledge: knowledge that was dearly valued considering that Slater would be entitled to half of the profits of the business and would own half of the machinery, after paying for his half of the manufacturing costs out of his future profits.

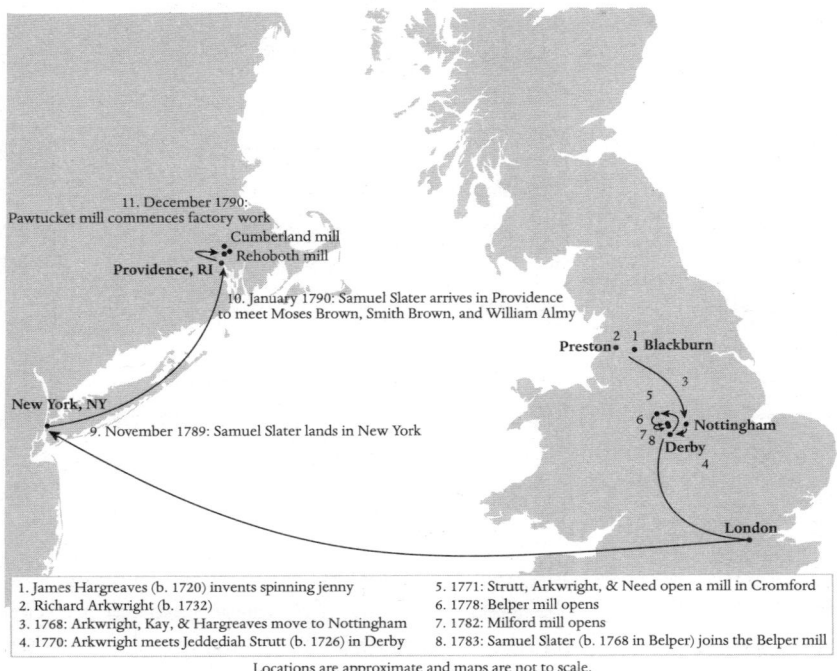

11. December 1790:
Pawtucket mill commences factory work
Cumberland mill
Rehoboth mill
**Providence, RI**
10. January 1790: Samuel Slater arrives in Providence
to meet Moses Brown, Smith Brown, and William Almy

Preston 2  1 Blackburn
5      3

6 ⟳ Nottingham
7 8 Derby
4

**New York, NY**
9. November 1789: Samuel Slater lands in New York

**London**

| | |
|---|---|
| 1. James Hargreaves (b. 1720) invents spinning jenny | 5. 1771: Strutt, Arkwright, & Need open a mill in Cromford |
| 2. Richard Arkwright (b. 1732) | 6. 1778: Belper mill opens |
| 3. 1768: Arkwright, Kay, & Hargreaves move to Nottingham | 7. 1782: Milford mill opens |
| 4. 1770: Arkwright meets Jeddediah Strutt (b. 1726) in Derby | 8. 1783: Samuel Slater (b. 1768 in Belper) joins the Belper mill |

Locations are approximate and maps are not to scale.

*Figure 12: Map summarizing the journey of knowledge in Samuel Slater's story.*

But it was not all smooth sailing. The first attempt to card cotton failed. After close examination, the team determined that the inclination of the carding teeth was not quite right and needed a readjustment. Eventually, the enterprise succeeded, producing its first yarn in October of 1790, and commencing factory work on December 20, 1790; less than a year after Slater arrived in Rhode Island to meet Brown and Almy for the first time.

For us, the moral of the story is that some forms of knowledge don't travel well. Slater was working at the cutting edge of eighteenth-century technology. A technology which people in the United States had been desperately trying to reproduce. But manufacturing, or more precisely, water-powered cotton-spinning, arrived in the US not through an instruction manual or as an abstract idea, but inside the brain of a gutsy twenty-one-year-old born not far from where Jedediah Strutt erected one of his mills. Through a circuitous journey, knowledge that started in Blackburn, Preston, and Warrington made it to Nottingham and then to the small town of Belper, where a brilliant teenager absorbed it, and, in an act of treason, brought it to the United States.

But once the technology made it to New England its diffusion once again became local. Brown, Almy, and Slater's first dedicated cotton mill in Pawtucket was erected next to the falls connecting the Seekonk and Blackstone Rivers, in what today would be considered a suburb of Providence, Rhode Island. The next mills were built not far from the first one. One in Rehoboth, Massachusetts, on the opposite side of the river, and another one in Robin Hollow, about 2 to 3 miles north of the first mill. Ten years after the Rehoboth mill, the tiny state of Rhode Island had sixteen mills in operation, with seven more on the way. The mills contained between 13–14,000 spindles and were processing about 12,000 lb of cotton a week.[74] Slater did for the US what Arkwright and Strutt did for the English Midlands. And he could do so because in a deep sense he came from 'the future,' a future where the secrets of cotton spinning were diffusing and competition was becoming increasingly tough. Like the rabbit spreading wildly through the wilderness of Australia, Slater was an alien species benefiting from an environment that had yet to adapt to his genius. But his ability to carry knowledge was not an anomaly. For the last forty years, scholars in the field of innovation have been using statistics to formalize what Slater intuited as a teen: that knowledge travels in brains, benefits from social networks, and struggles to travel far.

★

Two hundred years after Slater changed the history of New England, not far from Pawtucket, a young economist by the name of Adam Jaffe became interested in the diffusion of knowledge. Adam was inspired by his academic advisor, the Lithuanian economist and Harvard professor Zvi Griliches, who had authored a pioneering study on the diffusion of innovations focused on the adoption of hybrid corn.[75] The problem was that, in the 1980s and 1990s, economics was dominated by mathematical models like those advanced by Paul Romer which were a bit too abstract to be mapped onto the real world. Adam was among those looking to develop the empirical counterpart.

One of the most famous economic modelers was Paul Krugman. Krugman's models focused on the spatial distribution of economic activities, what scholars call economic geography. He understood that knowledge shaped the geography of economic activities, but he was skeptical about anyone's ability to make empirical progress in the field. This led to an infamous quote stating that knowledge left 'no paper trail by which they may be measured or tracked.'[76]

This quote motivated Adam, a rising star in Harvard's economics department.* He knew that a paper trail existed, but it was hard to trace using computers running on punched cards. By 1989, Adam had already produced important studies showing that corporate

---

* I asked Adam personally if he had been partly motivated by this quote during an event we both attended in Nice, France during the summer of 2022. His answer was a resounding yes.

patent rates benefited from proximity to local universities[77] and firms.[20] But he wanted more evidence. So, after completing his PhD and joining the Harvard faculty, he teamed up with Rebecca Henderson, who we know for her work on architectural knowledge, and Manuel Trajtenberg, an Argentinian-Israeli economist, to track knowledge flows using patent citations.

In a patent, inventors are required to cite the current state of the art to demonstrate the novelty of their contributions and put it in context. These citations represent an important paper trail documenting the lineage of technology. During the review process, patent examiners usually add citations, increasing the observed linkages among technologies. Using this data, the economists found that a patent was about three to four times more likely to be cited by patents from the same city compared to other patents of similar age and technology.[21] This was evidence that knowledge is more likely to travel shorter than longer distances, just as in Slater's story, where knowledge moved quickly among nearby locations, such as Pawtucket and Rehoboth, but only crossed the Atlantic once.

Jaffe, Henderson, and Trajtenberg's 1993 contribution became a flagship for the empirical study of the geography of innovation. It was soon followed by others.

David Audretsch and Maryann Feldman were another duo of economists interested in the geography of innovation. They wanted to know if innovations clustered in space more than other types of economic activities. In 1996, they published a study showing exactly that.[16] For example, they showed that in 1982 42 percent of all innovations in the computer industry came from California and that 39 percent of all pharmaceutical innovations came from New Jersey. Using statistical models they could also show that this concentration was higher in activities where new knowledge played a more important role. But like Adam Jaffe and his colleagues, Audretsch and Feldman did not have a mechanism they could use to explain these knowledge flows or agglomerations.

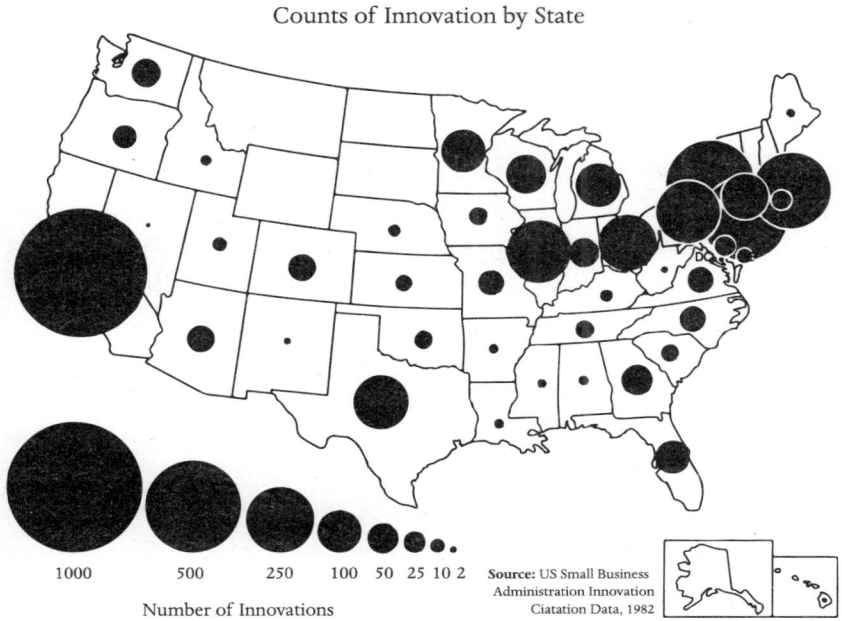

Figure 13: *Innovations by state. From Audretsch and Feldman.*[16]

Unlike the diffusion of a column of smoke, which can be explained by factors such as the direction of the wind or the Brownian motion of gas particles, the diffusion of knowledge has a more tenuous connection to geography. After all, knowledge is not really in a place. It does not diffuse through a continuous fluid, like the air, but through the discrete units we call people. So, the mechanisms explaining the geographic diffusion of knowledge must be reducible to something about us.

The two obvious candidates are inventor mobility, as we saw in the case of Slater, and social networks. Networks were a particularly hot topic in the early 2000s. The internet was taking the world by storm and scholars were grabbing any data they could to map and model networks.[78,79] Innovation scholars were not oblivious to this trend. Some of them began generating data on the mobility patterns and professional networks of inventors.

When an inventor moves into a new city they leave behind an enduring social network.[80] Their former friends and colleagues keep in touch with them, and more importantly, continue to cite them. Ajay Agrawal, Iain Cockburn, and John McHale used this quasi-experimental design to rule out the effects of 'pure' geographic distance. For instance, an inventor moving from Seattle to Austin continues to receive citations from inventors in Seattle but does not experience an increase in citations from cities at a similar distance from Austin, such as Boston.

Networks, in fact, explained a lot of the effects that a decade earlier had been attributed to geography. Stefano Breschi and Francesco Lissoni reconstructed the network of Italian inventors by looking at those who had worked together in a patent.[81,82] They showed that the odds of a patent citing another one was 150 times larger when the inventors had a history of direct collaboration, and sixteen times larger when they shared at least one common co-inventor. This idea was reinforced by Jasjit Singh, who also used a co-inventor network to show that patent citations followed the links of these collaborations.[83] It was networks, not space, that played a role. Space appears to play a role because social and professional networks are embedded in space.*

Again, a consensus begun to emerge. Knowledge flows were not constrained by geography but by the links of the social networks connecting inventors, craftsmen, and entrepreneurs; networks like those connecting Arkwright, Strutt, and Slater.

But there is still something unsatisfying about the social network

---

* These effects are related to the work of the sociologist Mark Granovetter on embeddedness. Granovetter has shown that many economic interactions, such as getting a job, are explained largely by the pre-existence of social networks created for non-commercial reasons (e.g. friends from school, church, etc.). I am not describing Granovetter's work here because I dedicate almost an entire chapter to his work in *Why Information Grows*.[84] If you are interested in Granovetter's work on embeddedness I would recommend his book *Getting a Job*,[85] his papers on the strength of weak ties[86] and embeddedness,[87] or chapter eight of *Why Information Grows*.

explanation. After all, social and professional relationships come in many shapes and forms. A relationship among peers, such as Strutt and Arkwright, is not the same as one between a master and an apprentice, like the relationship between Strutt and Slater. Does knowledge flow along any link? Or are some links more important than others?

One way to explore this question is using data on scholars who migrated forcefully to the United States in the 1930s.[88] The economists Petra Moser, Alessandra Voena, and Fabian Waldinger, used a dataset[89,90] just like that to identify twenty-six Jewish chemists who in the 1930s were unfairly dismissed from universities in Germany and Austria. Using patent data, they were able to find these chemists once they became active in the United States. Not surprisingly, the United States began to produce more patents in the fields where these chemists had expertise.* But what's more important is that the scholars could use this data to identify the links that facilitated knowledge diffusion. What they found was that most spillovers did not involve other chemists of a similar age and trajectory – the peers of the immigrants. The adopters were new entrants to the field. Their students. The knowledge brought to America by these chemists did not became part of the American chemistry

---

* A more recent version of a similar migratory event was the fall of the Soviet Union in the early 1990s. It resulted in a massive exodus of scientists and mathematicians who transformed the scientific landscape of Europe and the United States. This exodus was explored by Ina Ganguli, a professor of economics at the University of Massachusetts in Amherst, who found that, compared to similar papers from non-moving scientists, the papers of immigrant Soviet scientists received more citations from US scholars.[91] By moving to the United States, the Soviet scientists and mathematicians rewired their professional networks and the diffusion of their knowledge. George Borjas and Kirk Doran, respectively at the Harvard Kennedy School and the University of Notre Dame, focused on emigrating Soviet mathematicians, but found that one of the consequences of their arrival was to occupy some of the niches of local mathematicians, who in response to the sudden influx began working in different branches of mathematics.[92]

repertoire by influencing other senior chemists, but through acts of mentorship.[*][†]

But there are also other ways to differentiate among the links connecting people in a network. People's ability to influence others can depend on a person's perceived status, their ability to demonstrate skill, or how similar they are to others. In *The Secret of Our Success*, the anthropologist Joseph Henrich describes multiple studies exploring these mechanisms.[94]

One study that focused on prestige had preschoolers watch a video where two people played differently with the same toy. The videos included a couple of bystanders who provided a 'prestige cue' by focusing their attention on only one of the two players. When the preschoolers had a chance to use these toys later, they were over ten times more likely to use the toy in the same way than the player that had been paid attention to.[95] In a similar study focused on skill, children shifted their food choices to match those of other children who were good at solving puzzles.[94] These learning dynamics are also shaped by people's similarity. In the case of infants, it is well known that infants are more likely to learn from others who are of similar age to them.[96] In the case of college students, studies leveraging data on the random assignment of students to different instructors have shown that academic performance improves when a teacher matches the race of the students.[97] Social and professional

---

* This fact was actually anticipated in an earlier paper from Waldinger showing that faculty quality increases the probability that a student will publish their dissertation in a top journal and become a full professor.[90]

† It is worth noting that this is not just a story about science or chemistry. Mentorship has also been identified as a key channel for knowledge diffusion in the arts. A recent study by Karol J. Borowiecki, a professor at the University of Southern Denmark, focused on the works of composers.[93] Borowiecki looked at how often composers used the same subsequences of notes as a proxy for influence. He found that students wrote compositions that were significantly more similar to those of their teachers, and that this similarity remained significant for the better part of twenty years.

networks play a role in the diffusion of knowledge, but it is important to keep in mind that not all links are created equal.

While these studies are great in many ways, they also suffer from some limitations. One key drawback is that many of them can be hard to generalize to the 'real world,' because they are based on controlled environments – in the case of experiments – or impersonal administrative records – in the case of observational studies. For studies focused on patent citations, these records can be quite constraining since only a small fraction of the population works on patentable innovations. Even among those that do, propensity to patent is one of the things that people learn from their advisors.[98] Luckily, there are scholars who have also looked at the flow of knowledge using qualitative methods. These methods can capture some of the things that are hard to pin down using experiments or administrative data.

*

Transversely Excited Atmospheric or TEA lasers were invented in the 1960s in a defense research facility in Quebec by the physicist Jacques Beaulieu. When the news of the invention came out, other labs tried to build their own TEA lasers. But this was not an easy process. In what is now a classic 1974 study, the sociologist Harry Collins looked at the knowledge flows connecting teams working on TEA laser technology.[99] Collins had the intuition that the knowledge flows underlying these highly innovative activities must involve the movement of tacit knowledge (knowledge that is hard to codify). So, instead of focusing on the bibliographic record, he decided to interview the scientists working in labs focused on TEA laser technology.

By talking directly with researchers, Collins learned about interactions that were absent in the bibliographic record. That's how he was able to confirm that no one to whom he had spoken had 'succeeded in building a TEA laser using written sources (including preprints and internal reports) as the sole source of information.' This was in part because of the incomplete nature of scientific publications. As one of his interviewees admitted, publications are 'enough to show that you've done [something], but never enough to enable anyone else to do it.' This might explain part of the phenomena, but it is also a cynical statement that is hard to reconcile with the fact that these teams did help each other in person. Harris found smoking-gun evidence for the need for real interactions to transfer tacit knowledge, anticipating what Moser, Voena, and Waldinger would conclude decades later. Collins concluded that the transfer of complex and partly tacit knowledge requires apprenticeships and face-to-face interactions.

But there is something that Collins does have in common with many of the economists who studied this phenomenon quantitatively. His work still focuses on technical knowledge produced by scientists and inventors. So, to generalize these findings even further, we need examples of knowledge diffusion involving less formal settings. One example is a recent study that starts with the 'fall' – or 'liberation' – of Saigon in 1975.

★

In April, 1975, The United States lost its military control over Saigon in an event that unfolded quicker than many people expected. This led to an exodus of hundreds of thousands of Vietnamese who escaped in fishing boats and rafts.[100] In response to the unfolding humanitarian crisis, the United States created an inter-agency task force to relocate more than 140,000 refugees. These refugees had knowledge about Vietnam but were more likely to be small business owners and employees than elite scientists and inventors. With the support of many volunteer organizations, the US was able to resettle 130,000 refugees by Christmas of 1975. The resettling was hectic, with the task force releasing – on average – 4,000 refugees a week.* This meant that the refugees had to accept destinations chosen for them based on the availability and contacts of the volunteer organizations.

This quasi-experiment was exploited by Christopher Parsons and Pierre-Louis Vézina to explore whether these refugee assignments

---

*Between its establishment in May and December 20, 1975.

impacted future trade flows between the United States and Vietnam.[100] Following the war, the United States placed a trade embargo on Vietnam that ended in 1995. The end of the embargo opened the gates for subsequent trade, allowing the economists to study whether these new trade flows were partially explained by the 1975 relocation of Vietnamese refugees. Parsons and Vézina found that trade between the US and Vietnam was slightly larger in the states that had received more refugees. Even though twenty years had passed, trade between Vietnam and the United States still showed traces of people who had been distributed haphazardly across America.

*

More than 230 years ago Samuel Slater embarked on a one-way trip to the United States. His goal was to import knowledge that he had acquired in England as a teen. Looking back into his story we cannot help but think that the clever man from Belper understood things about knowledge and migration that scholars have only more recently been able to formalize. Slater understood that knowledge struggled to travel long distances, making him a unique and valuable asset in the young republic. He also understood that the knowledge he had was valuable enough to provide him with enormous leverage in negotiations involving powerful actors. But Slater also knew that his uniqueness was not going to last. He had seen the spread of water-powered cotton-spinning technology in England, and he knew that sooner or later the same would happen in the US. While he fought for the secrecy of his early operation, he also did not linger in the industry forever. By the late 1810s Slater retired from the business he had created, a business that in the subsequent years grew increasingly competitive.

Just as cotton mills blossomed around Pawtucket following Slater's arrival, so did studies on the geography of knowledge following the early papers on patent citations. But while these were pioneering studies, they were not completely unprecedented. Rich qualitative work, such as the TEA laser work of Collins or studies on the exodus of French Huguenots, anticipated many of the results that were later formalized by this quantitative literature.[101,102] So why do economists obsess about studying forced or unexpected migrations, like Vietnamese refugees,[100] Jewish-German chemists,[88-90] or Russian scientists?[91,92] Wouldn't it be easier to simply look at migration flows?

The reason is that migration flows conflate multiple mechanisms. When migration is not forced, and migrants choose whether to move or not, their choices may simply reflect personal characteristics, such as an entrepreneurial spirit or skill. This self-selection may explain outcomes such as patenting activity without the need for knowledge diffusion. Also, when migration is not forced, a migrant's choice of location may simply reflect the overall desirability of a destination. For many people, relocating to Paris or New York can be in principle more attractive than moving to Tripoli or Hanoi. That could mean that migrants are not contributing to knowledge creation but simply choosing places that are predisposed to produce knowledge in their respective areas of expertise. To really isolate the effects of knowledge flows, scholars must look for situations that get as close as possible to an alien abduction: a situation in which migrants are forced to leave at short notice and have little agency about where they go next. Forced migrations get close to this, albeit imperfectly, since even in the direst situations migrants have some choice of place. After all, the Jewish chemists we met above did not end up in Papua New Guinea, South Africa, or Chile, but in good universities in the United States. This is because, much like the dispersal of seeds, the fertility of a knowledge flow depends on the ground on which it 'falls.' This effect is known as 'absorptive capacity,' a key concept introduced by Wesley Cohen and Daniel Levinthal in the '80s and '90s.[103,104] The idea is that firms invest in R&D not only to generate new knowledge but to enhance their ability to absorb it. This concept is well illustrated in a legendary story about Masaru Ibuka, an engineer and one of the two co-founders of Sony.

★

Even though his father passed away when he was only two, Masaru inherited his knack for engineering. Masaru Ibuka was born in Nikko, a small city north of Tokyo in 1908.* As a second-grader, he enjoyed playing with an Erector Set, and in high school he became a ham radio operator. Ibuka studied mechanical engineering at Waseda University, and for his graduation project created a device that could transmit a signal as far as a mile and a half away by controlling the intensity of a light. His 1930 'light telephone' earned him a reputation as a student inventor and brought him to the 1933 World Fair in Paris where he earned the Gold Prize.

But even after making headlines in Japanese newspapers, Ibuka failed to land his dream job at the Shibaura Electronics Company (known today as Toshiba). He had a restless and creative mind that made it hard for him to follow rules. He was an inventor at heart, driven by a strong internal motivation, a quality that was valued by Taiji Uemura, the entrepreneur and co-founder of Photo Chemical Laboratories.

At Uemura's company, Ibuka's talents flourished. He was allowed to pursue his own research, and in return completed several jobs for Uemura. Together, they eventually founded Japan Measuring Instruments, a company where Ibuka worked as senior managing director from 1940 to 1945.

After the war, Ibuka returned to Tokyo as a veteran engineer, ready to become a founder. He dreamed of creating a workplace where engineers could follow their creative passions, and together with Akio Morita, a fellow wartime researcher, founded the Tokyo Telecommunications Engineering Corporation (Totsuko), a company that would become world-famous a decade later as Sony (a name they adopted in 1958).

The early days of Sony were a struggle for survival. They produced simple wares, like an electric rice cooker and a crude toaster

---

* The first pages of this chapter borrow extensively from John Nathan's excellent book on the history of Sony: *Sony*. (Houghton Mifflin Harcourt, 2001).

oven. But Ibuka was an inventor with more than a decade of experience who wanted more.

In June of 1949, he got a break while visiting the offices of Japan's national broadcasting service, NHK. The United States had taken over NHK's facilities. Next door to them, they had installed the Office of Civil and Information Education, which had one of the first tape recorders in Japan. Ibuka insisted on seeing the American-made machine and was taken aback by the quality of its playback. He observed the machine meticulously, concluding that the tape was moving at about 19 cm per second. The entrepreneur in him immediately understood the potential of the technology, but the engineer in him had to first figure out how to manufacture that mysterious brown tape.

From books and manuals his team understood that the tapes involved some form of magnetic coating. They also learned that they could produce magnetic powder by heating oxalic ferrite until it turned into ferric oxide. But that was easier said than done. Oxalic ferrite was hard to get. Eventually, they managed to obtain some from a pharmaceutical wholesaler in the black market of Kenda. Now they had to learn how to 'cook it.'

One of Ibuka's closest collaborators, Nobutoshi Kihara, learned how to toast the ferrite using a frying pan to the exact point it was brown, not black. He then combined the brown substance with shellac, a tree bug resin, to produce a magnetic coating. But the coating was only a part of the problem. To make a tape, the engineers needed to find a proper substrate. Cellophane stretched too quickly, distorting the sound. Paper was no good either, since it snapped under the tension of the machine. Eventually, they learned about paper strengthened with hemp fibers through a cousin of Akio Morita at the Osaka's Honshu Paper Company. Using expensive brushes made of badger hair, they could now coat the hemp-strengthened paper with their coating to produce the 'magic' tapes. Even though their tapes were of a lower sound quality than those made in America, they perfectly illustrate how a team's past research experience contributes not only to their ability to innovate but to their ability to absorb knowledge. If figuring out how to produce magnetic tapes with a frying pan, strips of hemp-reinforced paper, and a badger-hair brush is not a great example of absorptive capacity, I don't know what is.

Absorptive capacity is a key concept that helps us understand the receiving end of knowledge diffusion. This capacity can be essential for the diffusion of life-saving innovations. Consider the speed with which heart transplant technology spread in the late 1960s.

The first successful human-to-human heart transplant was conducted in South Africa, on December 3, 1967 by a team led by Dr. Christiaan Barnard. On January 6, 1968 the feat was reproduced in California, at Stanford's University Hospital. By the end of that year, more than 100 heart transplants had been completed. These transplants were not performed by random surgical teams, but by experienced teams of surgeons who were ready to absorb that knowledge.

Samuel Slater intuitively understood the idea of absorptive capacity. He did not embark on a solitary journey to build a cotton-spinning mill just anywhere in the world, but searched for people in the United States who were already embarked in that venture. And that's yet another reason why Samuel Slater's story is so interesting. It is a perfect storm where a smart entrepreneur self-selects to migrate, choosing the right place at the right time. That perfect storm resulted in much more than a measurable increase in patent citations, but on a turning point for the industrial history of the United States.

Slater's perfect storm is also a great example of key differences between research and policy. Scholars work hard to isolate effects by separating correlated factors. They want to isolate, for instance, the entrepreneurial knack of migrant inventors from the knowledge they carry. When it comes to policy, however, the fact that these factors are correlated can be a good thing. After all, countries benefit from attracting entrepreneurial migrants because they are both high-skilled and talented risk-takers.

In fact, talented migrants are not an exception. In *The Gift of*

*Global Talent,*[105] William Kerr provides numerous examples of the skill bias of migration. Migration is actually more common for scientists, inventors, and entrepreneurs, a fact that historically has benefited countries that are attractive for those who are skilled. Half of the founders of unicorn companies in the United States are immigrants, and about 40 percent of Fortune 500 companies in the United States were founded by immigrants or their children. The propensity for a person to migrate is also known to increase dramatically with education. About 1 percent of people with less than a high-school degree migrate, but about 5 percent of those with a college degree do. This propensity is more pronounced at the tails of the talent distribution. Ten percent of inventors with at least one patent are migrants, and so are 60 percent of the people who since 1970 have won a Nobel Prize. But this is not solely an American story. It is a story of advanced economies that are attractive to migrants. In fact, 15 percent of Swiss, Canadian, and Dutch patents granted by the USPTO involve migrant inventors.[106]

Together, these stories tell us that we should not think of skilled migrants as bringing an undifferentiated ability to work. After all, this could lead us to overestimate the negative consequences of migration. In a world where knowledge is fungible, meaning that units of knowledge are easily interchangeable, immigration will increase competition, lower wages, and raise unemployment across all sectors. But we don't live in that world. We live in a world where knowledge is non-fungible. So, we should expect labor displacement effects to be industry-specific. In a world where knowledge is non-fungible, we must consider the possibility that skilled migrants bring knowledge that is new and complementary. Knowledge that adds a new 'letter' to an economy's alphabet.

These stories also remind us that trying to stimulate knowledge production by building remote cities of knowledge, such as Neom or Yachay, can be an obtuse idea. There are many reasons why modern-day Samuel Slaters do not choose cities such as Neom or Yachay. They flock instead to places like New York, San Francisco, Tokyo, or Paris. The idea of building a city of knowledge in a remote location is at odds with the principles that shape the flows of knowledge, since skilled migrants are selective and often have many options.

Our understanding of knowledge diffusion is still incomplete, but it is safe to say that scholars agree on many key issues. The idea that the flow of knowledge decays with distance is not controversial, and neither is the fact that distance in a social network explains much of this spatial decay. It is also widely accepted that migration represents an important mechanism for the creation of the long-range connections that stimulate international knowledge flows. We can also say that part of what makes the flow of knowledge difficult is its tacit nature, as illustrated by Collins' study on the TEA laser. And we can also say that knowledge flows are harder to execute when knowledge is more difficult to absorb or more 'complex.'[107] We also know that apprenticeship and mentorships are a key knowledge-transfer mechanism, which is well illustrated in both the story of Slater and that of the chemists who forcefully migrated to the United States. Finally, the idea of absorptive capacity is also a well-accepted concept that reminds us that knowledge doesn't simply flow outward, but needs a fertile ground on which grow.

Today, research efforts focused on the flow of knowledge are no longer uncommon. But even before a modern paper trail existed, there was clear evidence of this principle. Slater might not have left the paper trail that Krugman wanted, but he did leave his mark in stone and timber: a trail of water-powered mills that were key for the industrialization of the United States.

# 7.

# The Principle of Relatedness

During the second year of my PhD, my advisor László Barabási – a well-known physicist and network scientist – asked me to visit an economist he had just met for lunch at a restaurant. The economist was Ricardo Hausmann, a professor of practice at the Harvard Kennedy School. Ricardo, who at some point in his life had wanted to study physics, had recently read László's book *Linked*, and was inspired by his prose and ideas. During lunch Ricardo asked László if he knew anyone who could help him make sense of big data.

When I started my PhD in 2004, I told László that I wanted to combine network science and economics. But László, who understood how hostile some disciplines can be to outsiders, told me that if I wanted to walk that path, I had to find the support of another advisor. This serendipitous connection was my chance to explore that opportunity.

After a brief exchange of emails, I headed to the Harvard Kennedy School to meet Ricardo. Our first meeting was on March 24, 2006, at 10 a.m. Ricardo's office was located on the fourth floor of the Rubenstein building, overlooking the JFK park and the Charles River. The office was divided into two smaller rooms, one containing a large bookshelf and his desk, the other a small rectangular table next to a whiteboard. I sat at the far end of that table, across from Ricardo and Bailey Klinger, a Canadian PhD student with a ton of 'rizz.'

Without delay, Ricardo started pitching me his idea. Countries, he said, were collections of private firms, or monkeys that lived in a forest. At the beginning, this sounded odd, but it is a clever way of thinking about the economic landscape.

The problem was that the trees were not all equal. Some trees had plenty of fruit while others were less productive. Moreover, the forest was patchy, with some parts being dense, full of trees with plenty of fruit, while other were balder, with trees far from each other. Of course, all countries wanted to have their monkeys – or firms – in the best parts of the forest, but that was hard to achieve if your monkeys were stuck up trees that were far from the center.

Ricardo and Bailey wanted to map the forest and were off to a great start. Using international trade data, Bailey had calculated a measure distance between each pair of trees, or products. But that's where they got stuck. They didn't know how to go from that to a map of the forest they could use to explore where the monkeys were located. That was my job. After the meeting, Bailey invited me to his office where I copied the international trade data they had onto an external hard drive. We agreed to meet again in a few weeks.

That spring I was working at the Dana–Farber Cancer Institute, a research and cancer treatment center affiliated with Harvard Medical School. The lab was led by Marc Vidal, a Belgian geneticist and systems biologist who, like the expert chef of a Michelin-star restaurant, commanded his lab with wit, humor, and wrath.

Back then, networks were a hot new topic in biology, but they were not yet a very popular idea in economics. Without having planned for it, I was suddenly becoming a bridge between two disciplines: a physicist doing a PhD on network science at a lab dedicated to mapping biological networks. I soon realized that the best way to represent 'the forest' was not to use a smooth landscape, like the rolling hills of western Massachusetts, but a network, like the ones we were using in Marc's lab to describe genetic interactions. The difference was profound. Unlike in a real forest where the ground is two-dimensional and pairs of trees that are close to each other are close to the same other trees, in the space of products distances are relational. What makes a product like natural gas similar to drilling machines, for instance, is very different from what makes drilling machines similar to steel tubes and pipes. Two 'trees' that are close to each other could be at very different distances from a third tree. To capture the geography of products we needed to abandon the spatial analogy. We were no longer looking at a regular lattice, but at a discrete cognitive space connecting knowledge and ideas. The forest analogy was brilliant, but to scientifically formalize it we had to shed the metaphor and think in terms of networks.

After a few days I was able to produce a network representation of the forest using software designed to visualize genetic interactions.* It worked perfectly for mapping the landscape of products defined by trade. This was not a coincidence.

Molecular biology data involves thousands of unique genes and proteins, and therefore requires representations that can help us keep track of each of the elements involved and their patterns of interactions.[108,109] Like a big Shakespearean play where each character plays a unique and well-defined role, genes and proteins are non-interchangeable, or non-fungible, just like products and the knowledge they require. Trade data is also made of thousands of non-interchangeable and unique categories, such as hot-rolled iron, calf skins, vulcanized rubber, and so forth. So, both proteins and products benefit from representations that preserve their identities. Networks provide that representation.

But even with the best tools available, making a network visualization was not easy. When done naïvely, network visualizations result in incomprehensible hairballs. I didn't want a hairball, but a well-designed 'subway map' showing clearly the connections between each pair of products. So, I kept the networks software running in the background of my little Acer laptop at all times, untangling the twisted web by hand, day after day, until no nodes overlapped, no links were too long or too short, and the angle of each vertex was just about right. After about a month I had the first version of what would later become a popular network. A network

* Cytoscape.

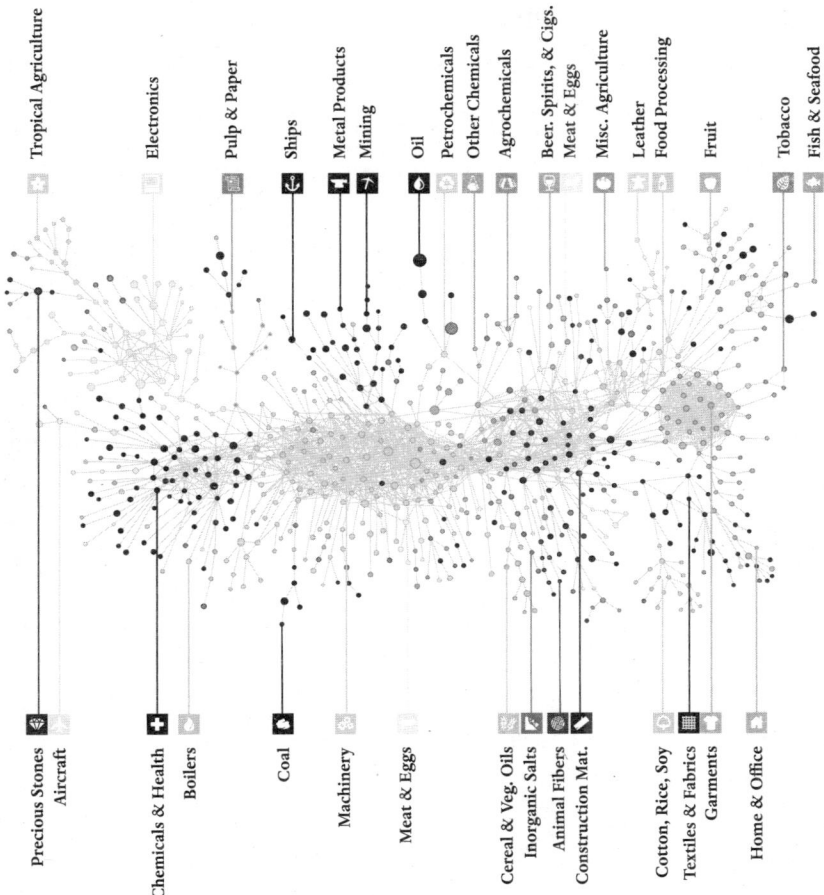

Figure 14: Network visualization of the product space. Each node here represents a product, and products are connected if they are likely to be exported by the same countries.

where I could visualize the movement of the metaphorical monkeys by highlighting the products a country specialized in using black squares. A detailed image of an economy that came not from a microscope or a telescope, but a 'datascope' that once properly oriented could be used to explore and explain decades of specialization patterns.

As we expected, the monkeys moved in a way that was highly constrained by the shape of the forest. With our datascope we could see how South Korea diversified during the 1970s, '80s, and '90s, as it entered several products in the electronics and machinery clusters. We could see Chile expanding from a few categories in fruits and fish to a larger variety of food exports. We could see Bangladesh's garment sector exploding as it began exporting dozens of unique types of apparel. We could see economic development not as a line on a chart where GDP moved up or down the *y*-axis, but as a differentiated process where thousands of sectors were present and patterns of specialization were unique. Not like an electrocardiogram, but a genetic screen. Like an embryo growing eyes, ears, and limbs, not just growing in size. We could finally understand the health of an economy using more than the weight of the baby.

This research turned into an academic paper that – in hindsight – turned out to be important because it moved the field into a relatively understudied aspect of knowledge diffusion.* Work on the diffusion of innovations historically focused on knowledge crossing oceans and mountains. These geographic efforts eliminate the need to consider the non-fungibility of knowledge by focusing on a tight domain: chemists begetting chemists and heart surgeons learning about heart transplants. These are examples of knowledge moving across geographies, not among activities.

But knowledge can also move among activities, and that second geography is the geography of the forest or the 'product space.' It is a complementary or dual 'geography' that is best represented by networks where pears are close to apples, guitars are close to banjos, and vaccines are close to antibiotics. It is not about knowledge moving from Boston to New York, but from Liberty shipbuilders

---

* There were actually a few papers in the field that did construct a measure of similarity between activities but not visualized as a network or used to explain international differences in economic development. Jaffe 1986[20] is a notable example. More recent examples are found in the work of Corinne Autant-Bernard.[110] At the time we were working on the product space paper I was completely unaware of this work, and to the best of my knowledge, so was Ricardo.

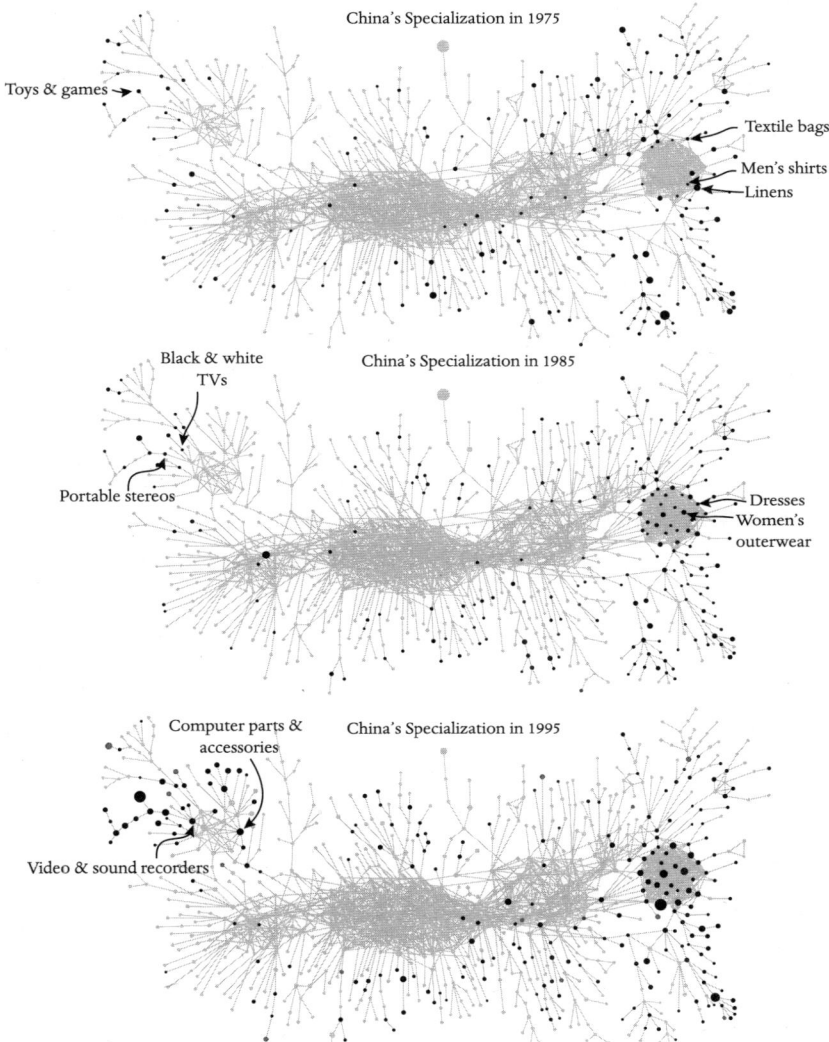

*Figure 15: Evolution of China's export structure as seen in the product space. Black nodes represent products China is specialized in. Light gray nodes represent nodes China is not specialized in.*

learning from aircraft manufacturers. This dual geography is important, because it allows us to incorporate the non-fungibility of knowledge into our models of international and regional development. It was the gateway to incorporating non-fungibility into the theory.

Back then, we did not frame our study in terms of the non-fungibility of knowledge. It would have probably been too abstract to engage a multidisciplinary audience. Instead, we motivated our work by focusing on what the product space did, help explain that economic development was difficult for economies in the peripheral regions of the 'forest.'

On July 27, 2007, our study was published in *Science*. This happens to be my maternal grandmother's birthday, and, sixteen years later, the day my beloved maternal grandfather Antonio – her husband of more than seventy years – passed away. But I digress. I remember receiving a call from László. He was having drinks with Marc, and they were starting to get loud. He shouted to Marc that I just got a paper accepted in *Science*: the most important multidisciplinary journal in the United States. Marc said something like 'No shit!' They knew better than me what that meant.

★

The Second World War devastated the business of the Piaggio family. Until then, they had been making airplanes. But after their manufacturing plants were bombed, and Italy was banned from manufacturing aircrafts, they needed a new way forward.

What the war did not destroy, however, was the brilliance of some of Italy's greatest aircraft engineers. Among them, we find Corradino D'Ascanio, the genius from Abruzzo who was to Italy what Theodore Wright was to the United States.

D'Ascanio was a prodigy.[111] He had started building hang gliders at the age of fifteen, three years after the Wright brother' famous first flight. Inspired by the age of aviation, D'Ascanio grew up to become a helicopter designer, building several 'vertical flying machines' that set world records for duration, distance, and altitude. But D'Ascanio's company crashed during the Great Depression, and in 1932 he began collaborating with the Piaggios as an expert on aircraft design and manufacturing.

During the following years their relationship grew, and so did the recognition for D'Ascanio's work, who was promoted to the rank of general during the Second World War. But the factory bombings also ended D'Ascanio's aeronautic dreams. In the forest analogy, the aircraft tree had been bombed, and the monkeys – which included the Piaggios and D'Ascanio – were forced to jump onto a new one.

Soon, D'Ascanio was approached by Ferdinando Innocenti, an entrepreneur and metal-tubing manufacturer. Innocenti wanted D'Ascanio to design an inexpensive scooter that was easy for men and women to ride and that would avoid getting their clothes dirty. D'Ascanio didn't particularly like motorcycles, but he took the job anyway, designing a vehicle with a rear-mounted engine that connected directly to the wheel, and a frame built on a ground-level shield that acted as a giant mudguard. Because the engine was in the back, mounting the scooter was easy, making it accessible to both men and women. But D'Ascanio wanted a pressed steel frame and Innocenti, in an attempt to build on his pre-war business, wanted a tubular frame. Eventually, the collaboration fell out, leaving unfinished what a few years later would become the Lambretta, an iconic Italian postwar scooter manufactured by Innocenti.

D'Ascanio returned to the Piaggios, who were also trying to develop a postwar scooter. Unhappy with their existing designs, the Piaggios welcomed back the prodigal son, who went on to design a motorcycle with a load-bearing body that acted as a mudguard. The scooter also had a cantilevered wheel mount, similar to the one used on airplanes,[111,112] facilitating removing and repairing the wheels. The Piaggio's scooter was sold as the Vespa, meaning 'wasp' in Italian, hitting the market ferociously in 1947 a little bit before the Lambretta.

Louis Pasteur once famously said that chance favors the prepared mind. While postwar Italy had a widespread need for affordable transportation, D'Ascanio and the Piaggios were not random entrepreneurs. They were in a strategic position to take advantage of such as opportunity. The Piaggios were ejected from their industry, left with nothing but their knowledge, knowledge of how to shape metal into planes and helicopters they used to build wasp-like scooters. But while the Vespa story reads like an anecdote, it was in fact part of a widespread global trend.

More than 9,000 km east of Italy, Japanese aircraft manufacturers were facing a similar conundrum.[113] Manufacturing facilities had been a key target for the American and Soviet militaries, forcing Japanese manufacturers to reinvent themselves after the war.

Aichi was a prominent Japanese warplane manufacturer. It learned how to manufacture planes during the years leading up to the Second World War by benefiting from a clause in the Treaty of Versailles. The clause required Allied inspectors to oversee German airplane manufacturers, and since Japan had been a member of the Allies during the First World War, it could participate as an inspector. During the interwar period, German aircraft manufacturers were still allowed to build civilian aircraft, a license that they used to continue designing bombers as 'postal aircraft' and fighters as 'acrobatic planes.' Aichi exploited this loophole to learn about warplane manufacturing from Heinkel in exchange for informing the company about upcoming inspections.

But after the Second World War, Aichi also had to find a new way forward. In April of 1947 it began producing a three-wheeled minicar: the Giant AA1.

Kawanishi is another example. The industrial group with factories near Kobe, Osaka, and Himeji specialized in aircraft for the navy. But as the Second World War was coming to an end, ninety-two B-29 bombers destroyed their Kobe factory. In 1946, the survivors started producing a civilian motorcycle at their Nishinomiya factory: the Pointer, sold later as the Lassie in the United States.

Kawasaki was also involved in warplane manufacture. The iconic company made nearly 2,000 Ki-48 'Lily' bombers and many well-known fighters. Their factories were also destroyed in 1945. But unlike other Japanese aircraft manufacturers, they managed to stay in the aircraft industry by making parts for American planes in the early 1950s when Japan was allowed to resume some activities in the aerospace industry.[114] In 1962, Kawasaki introduced its first motorcycle, the B8.

In Germany, Heinkel followed a similar fate. The manufacturer of fighters had helped pioneer jet propulsion as early as 1937, thanks to the work of Hans von Ohain, a young physicist who had earned his PhD at the University of Göttingen only four years earlier. Von Ohain's jet engine led to the first flyable jet plane, the Heinkel He 178, which took to the skies on August 27, 1939.[115] But after the war, Heinkel also went into the bicycle and motorcycle business.[116]

You should see a pattern here. After the Second World War, the Italian, German, and Japanese firms that were no longer allowed to manufacture airplanes jumped into the closest 'tree' they could: light vehicle manufacturing. The destruction of their factories, combined with the postwar aircraft manufacturing ban, forced these companies to jump into new products. But since the bombings did not destroy their knowledge, they all chose similar ones. No matter if the companies were Japanese, German, or Italian, they all left a trail of smoking-gun evidence of links in the product space.

Luckily, this is not the first time we have encountered this principle. When Richard Arkwright looked for a partner for his spinning-machine, he searched out John Kay, a watchmaker who could bring in related knowledge of precision metal manufacturing. Arkwright himself also had experience with hair, from his time as a barber and hair merchant. When Masaru Ibuka began working on magnetic tapes, he leveraged decades of experience in electronics manufacturing. When Edwin Land had to find a new postwar business for Polaroid, he used his knowledge of chemistry and light polarization to create instant film. All of these cases show that the diffusion of knowledge is not only constrained by the geography of rivers and mountains, or by social networks. They are also shaped by the landscape of knowledge. A landscape that is harder to see but which is just as real.

⋆

In 2009, at a conference at NYU, I met Otto Koppius, a jolly assistant professor in Business Analytics at the Rotterdam School of Management (RSM). A few months later, Otto invited me to the Netherlands to present my work. At a dive bar not far from a quiet canal I discovered he was a pinball wizard. He played the silver ball for more than half an hour with a single credit in the machine.

As is customary, Otto scheduled meetings between me and some of his colleagues. One of them was Frank Neffke, an assistant professor at the Erasmus School of Economics who was working on a remarkably similar problem. I sat with Frank in an empty meeting room where he showed me how he was using data from Sweden to build a network connecting industries that manufactured products in the same plants. Frank's industry space was – of course – another incarnation of the product space. But it nevertheless looked a bit different. Unlike the product space, which had a densely connected core and a frizzed hair periphery, the industry space was composed of clusters focused on the production of food, chemicals, fabrics, and metals. Like the product space, Frank's data was restricted to manufactured products, so it did not include doctors and lawyers, but connections among industries procuring items like toothbrushes, metal wire, and shirts from the same plants. Still, Frank's main result was remarkably similar to ours: the probability that a Swedish region would enter an industry increased when the industry was close to that region's 'monkeys.' The raw effect was again rather large, going from about a 2 percent chance of entering an industry in five years, when no 'monkeys' were nearby, to almost 20 percent when a region was brimming with 'Swedish monkeys.'*

Frank's study was published in 2011,[117] but one reason why it represents a turning point in our story is because it marks the beginning of a sequence of reproduction. During the following years, several scholars began using different datasets to reproduce and extend our 2007 finding. This included the work of Dieter Kogler in Dublin, who championed a couple of studies extending these ideas to the patenting activity of cities in the United States.[118,119] Miguel Guevara, a Ecuadorian graduate student who visited my lab in 2014, championed a related effort using data on academic publications.[120] This was not an easy task. Miguel and I were interested in individual scholars.† In Miguel's network, neurobiology and medicine were connected if the scholars who published articles in neurobiology also published articles in medicine. Going down to individuals was important to us because individuals are more constrained by their expertise than countries or universities. So, we should expect the links of a network built on individual level to be more representative of shared expertise. As we expected, Miguel found out that at the individual level the effect of relatedness was considerably larger than the one we observed for universities and countries. This supported the idea that relatedness was about shared knowledge.

---

* Here I am using the unconditional effect, as can be read directly from the chart, instead of an effect with controls. This is for simplicity of language and to illustrate the larger point, which is that the effect is not a blip, but a visible regularity that – as shown later in the chapter – is easy to reproduce.

† Getting the data for individuals was particularly challenging because many authors share common names, such as Juan Perez or John Smith. To circumvent this problem, Miguel collected data from Google Scholar, a service that provides an incentive for authors to curate their own data. Google Scholar doesn't make it easy for anyone to collect that data, so it took Miguel about a year to build the dataset we needed to connect a relatively well-disambiguated set of authors and papers. Using that data, we could finally connect academic fields based on the probability that an author had published in both of them.

As the study of relatedness* continued to heat up, people started to grab any data they could find to map a 'forest.' This transformed the once surprising effect began moving into something that become so easy to reproduce that the field began moving into a couple of different directions.

The first one was an effort to generalize the idea. In 2016 I began inviting colleagues to participate in a joint article making the case that what we were observing was not a collection of anecdotes, but a general principle; a reproducible law of economic geography that was valid across multiple scales and activities. True not only for countries and their exports, but for countries, cities, and regions, and for industries, products, patents, papers, and occupations. The colleagues included an eclectic mix of European economic geographers, such as Frank and Dieter, and mainstream economists, such as Edward Glaeser, who today heads the economics department at Harvard University. Eventually, we published the article at the Conference on Complex Systems, where I was invited to present my work as an invited speaker in 2018. Now the 'Principle of Relatedness' is the name used by specialists to talk about the highly reproducible and generalizable fact that metaphorical monkeys don't jump far from metaphorical trees.[121]

The second set of research went in the opposite direction: toward the particular. It involved efforts to unpack the principle of relatedness. Was relatedness a reflection of links along value chains, shared infrastructure, natural resources, or shared knowledge? And in the case of the last of these, was relatedness a reflection of knowledge specific to an industry, occupation, or location? Exploring these questions required having multiple measures of relatedness.

An early effort in this direction was work by Glenn Ellison from MIT, together with Edward Glaeser and William Kerr from Harvard. Ellison and Glaeser had developed a co-agglomeration or colocation index back in 1997,[122] but they had not used it to explain 'entries'

---

* This is the technical term we use to talk about how close a country or a region is to an activity.

or 'exits' (the technical phrases used to describe monkeys jumping onto or off a tree). In 2010, they followed up on this work by asking if labor or input–output relationships were better at explaining which industries collocate. They concluded that all of the channels contributed, but that input–output links played a special role.[123] This work was then expanded by Frank Neffke, together with Dario Diodato and Neave O'Cleary, who included a better measure of shared skills.[124] They found that, while value-chain relationships were important early in the twentieth century, their role had been steadily declining, concluding that modern patterns of co-agglomeration were better explained by pools of shared skills than value chains.

My team also became interested in unpacking relatedness, but instead of focusing on what explained collocations, we wanted to know what explained the success of new firms. This required data that was hard to get, but that we had thanks to a relationship I had forged a few years earlier with colleagues in Brazil.

In 2010, a week after I had joined MIT as an assistant professor, I received an email from Virgilio Almeida, a well-known computer scientist and professor at the Federal University of Minas Gerais (UFMG) in Brazil. Virgilio was part of a group of publicly minded academics who wanted me to visit Belo Horizonte to discuss my work. In Belo, Virgilio introduced me to Evaldo Vilela, a Brazilian entomologist who a few years later became the president of FAPEMIG: the scientific funding agency for the state of Minas Gerais. That trip started a relationship with Brazil that is still ongoing, and which has led to the creation of several online economic data observatories.

Our first project, Data Viva, was an unprecedented economic data observatory designed to visualize information on the exports, industries, occupations, and education for each of Brazil's 5,000+ municipalities.* To create Data Viva we needed detailed employment data, which in Brazil can be found in a dataset called RAIS.

---

* It was a paramount effort led by Alexander Simoes and David Landry, with whom I co-founded Datawheel, a company focused on the creation of public

RAIS is an amazingly detailed social information dataset where each row represents a person, and each column tells us where that person worked and what job they did. We would know, for instance, that the person in row 123 would be a twenty-seven-year-old college educated female working as an accountant for a car repair shop in the municipality of Contagem. Since we had this data for several years, we could also see where that person was working the year before or after. For instance, we would know if that person became a financial manager at an art museum in the city of Manaus. This allowed us to see how knowledge moved among occupations, industries, and locations. Like bees carrying pollen from one flower to the next.

The 'dirty work' of that study was done by Cristian Jara-Figueroa, an exceptionally talented Chilean PhD student who joined my team in 2014. Cristian wrangled that data to calculate two types of relatedness. One based on similar industries, and another one based on similar occupations. This gave us the ability to approximate three types of knowledge: knowledge that was specific to an industry, an occupation, or to a location.* In the case of worker 123, she would be carrying occupation-specific knowledge from Contagem to Manaus, since accounting and financial management are related occupations. But she would not be carrying much industry-specific knowledge, since car repair shops are not similar to art museums. She would also not be carrying much location-specific knowledge because Contagem and Manaus are far away from each other.† Location-specific knowledge is important, among other things, because it tells us if a worker has a local network of social and professional contacts.

---

data distribution systems. Alex did his master's with me at MIT, developing the Observatory of Economic Complexity (OEC) (oec.world) as part of his master's thesis and continued working with me on economic data visualization as soon as he graduated. The OEC helped convince our Brazilian colleagues that a platform was a better idea than a report.

* The knowledge accrued after having worked for a while in the same city.

† More than 2,000 km from each other.

We could now use these three estimates to see which one had the largest impact on the success of 'pioneer firms.' Pioneer firms are the first ones to operate in an industry that is not present in a location. They represent the simplest expression of industrial diversification. The first monkey to jump onto a new tree.

Putting the three measures of relatedness to compete in a 'horse race' revealed that industry-specific knowledge was better at explaining the growth and survival of pioneer firms.[125] Location-specific knowledge came in a distant second. Surprisingly, occupation-specific knowledge didn't seem to matter once we took the other two forms of relatedness into consideration.*

We then started working on how to unpack relatedness in the context of international trade.† When a country starts exporting a product, they do not export it to just any location. It is easier for a country to enter some destinations than others. This depends on whether they already export similar products to that market, whether they export the same product to a neighbor of that market, or whether one of their neighbors is already exporting that product to the target destination. This sounds intuitive, but being able to scale that intuition to millions of bilateral trade relationships meant that we could run models to estimate the potential of a country exporting a product to every country in the world. Measures that we could use to anticipate the 'inertia' of global trade.[126]

A third strand of research that emerged during the 2010s focused on unrelated diversification. This is the study of cases in which countries, cities, and regions 'violate' the principle of relatedness.

---

* To finalize the paper, Cristian used a technique called a Bartik instrument to show that this was not a mere correlation, but a robust, maybe even causal, effect. In a nutshell, a Bartik instrument captures the idea that in the presence of an industry-specific external shock, such as an international drop in prices forcing it to lay off workers, related local industries benefit from the excess of talented workers looking for a job.

† This was work conducted with three members of my lab at the time: Bogang Jun, a Korean economist and postdoc, Aamena Alshamsi, a computer scientist from Abu Dhabi, and Jian Gao, an applied mathematician from China.

By definition, unrelated diversification cannot be explained from the activities that are already present in a location, so it must involve knowledge coming from somewhere else. Not surprisingly, this is where the migrants from our previous chapter come in. Relatedness matters about 20 to 30 percent less for migrant inventors compared to local inventors.[127] Also, migrants are more likely to spark unrelated diversification back at home,[128] reducing the impact of relatedness by about 10 to 20 percent.

But there was one remaining frontier. Since all of the work I've mentioned uses data of relatively recent vintage, we cannot use it to generalize this principle to other time periods. Was the principle of relatedness present in Europe at the time of the Renaissance or during the Industrial Revolution? Or is it something that emerged more recently? Without historical data, we can only speculate.

For me, this was a long-standing question that I was finally able to explore when Philipp Koch, a brilliant economist from Austria, and Vienna's 2022 snooker champion, joined my lab in Toulouse to work on his PhD. Philipp and I share a passion for history, so our conversations quickly began to revolve around historical patterns of economic development. Luckily, we had some data at hand.

Back in 2013, my team and I had built Pantheon (pantheon. world),* an observatory of collective memory visualizing fine-grained data on dozens of thousands of biographies. With Philipp, we could use this information to study the cultural specialization of European cities going back centuries. The idea is that the places of birth and death of famous historical figures, such as painters, composers, and mathematicians, can tell us, for instance, that Vienna was specialized in music and Paris was a beacon for the visual arts.

Philipp used this data to explore questions about the diffusion of knowledge going back 1,000 years. Our data on migration was spotty, since we had to approximate migration using the places of birth and death of famous historical figures. Philipp tested

---

* With Amy Yu, Shahar Ronen, Kevin Hu, and later, with Alex Simoes and Cristian Jara-Figueroa.

the validity of this assumption using a random sample of 200 biographies, finding that places of death were an important place for more than 70 percent of biographies. He was also able to show that the probability that a city gives birth to a famous person who speciality in an activity increases if that city has received an excess number of migrants specialized in a century earlier. To reduce the risk of reverse causality, Philipp controlled for an enormous number of specific factors, allowing us to rule out hypotheses such as a new university attracting scientists or a new conservatory attracting musicians.* This was already an important contribution, but Philipp wanted to go further. So, he created three estimates of relatedness. One for a city's 'locals': people who were born and died in the same place; another one for a city's emigrants: people who left a city; and another for a city's immigrants: people born elsewhere who died in that place. The statistical models showed that migrants were the main contributors to the principle of relatedness. That is, migrant physicists not only contribute to the production of more physicists, but to the production of chemists, astronomers, and mathematicians. Surprisingly, the relatedness of locals didn't seem to matter after considering migrants.† Philipp was able to show that the principle of relatedness was a historical reality, a principle that had been shaping the diffusion of knowledge on our planet for centuries.

*

In the early 1990s, Joseph (Joe) Jacobson and his brother were back-packing Peru. While having dinner one day, they met a Peruvian teen who showed them a notebook full of engineering sketches. Joe was stunned by the teen's technical knowledge. How could someone

---

* Since these effects are specific to a city and a time period.
† This could be because the Pantheon data is a very selective sample composed mostly of highly talented migrants, ranging from very well-known cultural figures, like Leonardo da Vinci, to others who are less so, such as Israel Tsvaygenbaum.

from such a remote location know so much about engineering? The teen then showed Joe his collection of science and engineering magazines. This serendipitous encounter inspired Joe. He began to imagine a world where curious teens anywhere in the world no longer had to rely on a stack of magazines but could instead use a digital notebook that was constantly updated with fresh content.

Back in the United States, Joe, a native of Newton, Massachusetts, completed a bachelor's degree and PhD in physics, and eventually joined the MIT Media Lab as an assistant professor. There, he became a co-inventor of electronic ink. Joe's success as an inventor, however, was the result of multiple factors. Unlike the teen he met in Peru, he was born in an affluent American neighborhood and had the good fortune to attend Brown University for his undergraduate and MIT for his PhD. He also had access to great talent, including Barrett Comiskey and Jonathan Albert, two undergraduates who toiled endlessly in his lab until they figured out how to produce electronic ink. But as Joe told me personally, relatedness also played a factor in his story, because in the late '90s Joe's lab was only a short walk away from Polaroid, a company with related knowledge and specialized equipment that helped him and his team bring e-ink to life.

Making electronic ink for the first time required combining knowledge available in labs that were a few blocks away from each other, physically, socially, and cognitively. Labs that were present in Cambridge, Massachusetts, but not in Cusco, Peru. Like a modern D'Ascanio, Joe leveraged this local knowledge to create a product that was new to the world. Even at the cutting edge of technology, the accumulation and diffusion of knowledge follows principles. These principles form the second law. The law of knowledge diffusion.

## The Second Law: The Principles of Space

Knowledge diffusion is governed by two complementary principles: diffusion across geographies and social networks, and diffusion across activities or cognitive domains.

Knowledge diffuses more effectively within short distances because these tend to be connected by social and professional networks. This diffusion, however, is not amorphous. It tends to work more effectively from masters to their apprentices than among peers. Knowledge diffusion also depends critically on migration, whether voluntary or forced, as migration catalyzes the creation of bridges connecting geographic divides.

But there is a second, less obvious geography that also affects the diffusion of knowledge. This is a geography connecting related activities, such as similar products, industries, or technologies. These networks also constrain the diffusion of knowledge. This 'principle of relatedness' says that knowledge is more likely to take root in fields closely linked to existing expertise, as illustrated by the fate of some Second World War airplane manufacturers in Japan and Europe. This principle underscores that the transfer of knowledge among fields is not random either, but shaped by networks connecting fields, such as the product space or industry space.

Together, these principles reveal that knowledge flows are not only bound by physical or social proximity but also by the cognitive and structural connections between different domains of expertise.

# PART III

## The Principles of Value

# 8.

## *More Than Rules*

Bretton Woods is a recreational ski area next to the small town of Carroll, New Hampshire, nestled in the foothills of Mount Washington, about three hours north of Boston and about ninety minutes south of the Canadian border.

Most visitors come to Bretton Woods during the winter. But in July of 1944, the Mount Washington Hotel was busy hosting the more than 700 delegates who had come to participate in the United Nations Monetary and Financial Conference. The representatives came from all forty-four Allied nations, who had grown confident about their chances of winning the war. The goal of the conference was to prepare the reconstruction effort by establishing new organizations and designing a new system for international monetary exchange. For twenty-two days, delegates, including those of the caliber of John Maynard Keynes, deliberated intensely. On July 22, the conference concluded with the signing of the now-famous Bretton Woods agreement. The agreement established the US dollar as the linchpin currency of the international monetary system and kickstarted the creation of the International Monetary Fund (IMF) and the International Bank for Reconstruction and Development, the leading lending arm of today's World Bank.

Bretton Woods is a historical example of a gathering that led to the creation of some important formal institutions. It defined new rules of the game that governed international monetary exchanges for nearly thirty years.* But Bretton Woods is also a good starting

---

* Until August 15, 1971, when the convertibility of the US dollar to gold was suspended, making the dollar a fiat currency.

point to understand the evolution of the way in which we think about economic development.

European reconstruction was a top priority during the aftermath of the Second World War. The Western Allies recognized that rebuilding the continent's economy was essential for sustaining peace and countering the threat of communism. As a result, the postwar period saw substantial financial support directed toward European reconstruction.

European reconstruction was a chief concern of the newly minted World Bank. In March 1947, John McCloy, a former Wall Street lawyer who advised presidents from Roosevelt to Reagan, became the World's Bank second president.[129] About a month later, the bank approved its first reconstruction loan, a $250 million reconstruction credit to France. Using his Wall Street connections, McCloy helped the bank raise additional capital by officially entering the bond market in July of 1947. By the end of that summer, the bank approved a $195 million reconstruction loan to the Netherlands, a $40 million loan to Denmark, and a $12 million loan to Luxembourg. But McCloy understood this was not going to be enough. On March 4, 1948, he testified to the US Congress that 'in the opinion of the bank's staff, the proposed 6.8B USD . . . recommended by the administration' for the reconstruction of Europe were 'a rather tight fit.'[130] The point was that the effort needed to support the reconstruction of Europe had to be larger, even if this crowded out the bank's own lending program. In four years, the European Recovery Program, known popularly as the Marshall Plan, transferred over $13 billion (about $170 billion today) to several countries in Europe. Today, many people regard the Marshall Plan as a crucial component of postwar reconstruction. The fact is that some key European economies did bounce back quickly during the postwar period. France's GDP per capita, for instance, doubled between 1944 and 1950, and then again between 1950 and 1968.[131]

But the European Recovery Program did more than support the reconstruction of Europe. It captured the minds of the world by suggesting that development could be engineered through finance. In the case of the reconstruction of Europe, finance seemed to be a necessary and sufficient condition. After all, everyone agreed that European economies had been decimated by the war. If financial aid was enough to reconstruct the war-torn economies of France and Germany, why couldn't we use it to develop countries whose economies had not been devastated by war? Given a big enough push, economies in Latin America, Africa, and Asia could be moved into higher levels of development. If that big push was provided by a loan, development could even be engineered at a profit.

For decades the world tried this idea, but eventually people began to realize that, although finance might be necessary, it was not a sufficient condition. Did Europe perhaps have something that the developing world lacked, something that couldn't be easily bought? The dominant answer to this question became 'institutions.' Until the 1980s, most development loans did not require countries to undertake institutional reforms. Loans focused primarily on the development of industry and infrastructure projects.[132] But in the 1980s, that began to change. The basic idea was that institutions, which can be thought of as the rules of the game, or as the American economist Douglass North famously put it, as 'the humanly devised constraints that shape human interaction,'[133] generate the incentives that shape human behavior. This goes from traffic laws dictating how we drive, to constitutions defining how long a person can serve as a president. That made the practice of development that of finding and promoting the right set of institutions. Since the United States and Western Europe were believed to have those institutions, what development agencies needed to do was to transplant these institutions into the least successful nations.

During the following decades the idea of institutions reigned supreme. It was also an idea that was right for its time. By the late '80s, the 'natural experiments' of East and West Germany and North and South Korea had run their course. After the fall of the Berlin Wall and the collapse of the Soviet Union, the answer to the

Cold War's ideological battle was clear. Francis Fukuyama famously declared the end of history.[134] The academic literature was brimming with work supporting the key role played by institutions. In 1990, Douglass North published his magnum opus on institutions and economic performance, a book that has since received around 100,000 citations according to Google Scholar. He went on to receive the 1993 Nobel Memorial Prize in Economic Sciences for his work on economic and institutional change.[135] After decades, the Cold War and the debate on institutions were finally over. Democracy and capitalism were the winning combination. What developing countries needed to do was to adopt these institutions.*

As the years progressed, the evidence supporting the role of institutions in economic development grew increasingly strong. In 2004 the Harvard Kennedy School economist Dani Rodrik declared that, when it comes to determining income levels, 'the quality of institutions "trumps" everything else.'[137] For Rodrik, everything else meant two things: trade integration and geography.[138] To test his hypothesis, Rodrik, together with Arvind Subramanian and Francesco Trebbi, used data on the mortality rate of colonial settlers as a proxy for the initial quality of a country's institutions. This was an approach that came from work done by Daron Acemoglu, Simon Johnson, and James Robinson, which contributed to them receiving the 2024 Nobel Prize in Economics. The idea was that 'in places where Europeans faced high mortality rates, they could not settle and were more likely to set up extractive institutions' (institutions designed to extract resources from the many to the few).[139] So, data on the mortality of colonizers could be used as an instrument to measure the quality of a country's initial institutions and connect these to economic growth.

But not everyone was convinced. Not far from Rodrik's office,

* A key example of that recipe was the Washington Consensus, a list of ten policy recommendations published by the economist John Williamson. These recommendations included the adoption of practices such as fiscal discipline, trade liberalizations, and deregulation (among others).[136]

Edward Glaeser and Andrei Shleifer* questioned the emerging orthodoxy by insisting that knowledge also played a role.[140] Their argument was that 'Europeans who settled in the New World brought with them . . . their human capital.' They showed that the instruments used in the literature as measures of institutions were highly correlated with measures of human capital in the 1900s, invalidating the idea that settler mortality represents a measure that isolates the effect of institutions. Immigrants also come with knowledge, and the demand for institutions, therefore, can come from those working in knowledge-intense industries.

This idea agreed with a couple of things that were becoming increasingly hard to ignore in the 2000s. The first one was that economies that were 'star reformers' did not seem to reap the benefits promised by those touting institutions. This is an idea explored well in Matt Andrews' book *The Limits of Institutional Reform in Development*.[132] The second one was the rapid economic growth of China. For decades, China had been growing using an 'unorthodox set of institutions.'[141] While it is well known that China began transitioning to capitalism in 1978, with the reforms introduced by Deng Xiaoping, it is a country that has not yet embraced key Western institutions, such as liberal democracy. But what could this mean?[†] Maybe this meant that the diversity of 'right' institutions is larger than what scholars in the early '90s originally imagined? Maybe it meant that institutions were also a necessary but not sufficient condition?[‡] Maybe China had something beyond institutions

---

* Together with Rafael La Porta and Florencio López de Silanes.

† According to James Robinson, the lack of Western-style institutions in China means its growth will be 'short-lived.' At least, this is what he argued at a recent event in Bali, Indonesia at which both of us had a chance to speak. He compared China to the Soviet Union. I personally disagree with this assessment because, despite the growth of the Soviet Union during the middle of the twentieth century, it never had the technological leadership in advanced consumer products that China has today.

‡ As it often the case with powerful ideas, sometimes people put too much faith in them. This is not because the ideas are wrong, but because sometimes, there

and finance that star institutional reformers with a poor economic performance lacked? My controversial view is that, despite China's abysmal economic output in the late 1970s, it was a country that was still relatively rich in knowledge. The good news is that you don't need to take my word for it, for this is the story of Chen Chunxian.*

\*

Everyone who has studied the economy of China associates 1978 with Deng Xiaoping. But 1978 was also the year of Chen Chunxian's first visit to the United States.

Chen was a Chinese physicist and nuclear fission specialist. In 1978 he was part of a four-person delegation that visited Princeton's tokamak fusion reactor. Tokamak reactors use strong magnetic fields to confine a high-temperature plasma. They were invented in the Soviet Union during the early 1950s by Andrei Sakharov† and Igor Tamm.‡ Chen Chunxian, who studied physics from 1952 to 1958 in the Soviet Union, succeeded at developing the first Chinese tokamak reactor in 1974. This qualified him for the delegation that visited the United States in 1978.

Chen's team wanted to learn how Americans built tokamak

---

is more to a problem than what we can immediately see. Some of this faith was clearly exaggerated. In his book, *The Limits of Institutional Reform in Development*,[132] Matt Andrews writes that in 2003 some members of the international community proposed that institutional reform would achieve a stable, centralized, and technocratic administration in Afghanistan within seven years. I don't need to tell you that this is not how that story ended. Andrews' book provides an interesting argument about the limits of institutional reform, by showing that star reformers are not necessarily star performers. This provides evidence in favor of the idea that, like finance, institutions may be a necessary but not sufficient condition.

\* This section of the book builds on the story of Chen Chunxian as told in Ning Ken's excellent book, *Zhong Guan Village*.[142]

† A physicist who received the 1975 Nobel Peace Prize.

‡ A physicist who shared the 1958 Nobel Prize in Physics for the discovery of Cherenkov radiation.

reactors. But during his trip, Chen learned about more than physics. He was surprised to discover that much of the manufacturing of plasma physics equipment in the United States was done by relatively small companies, sometimes as small as a dozen people. This was counterintuitive for Chen, who thought that American fusion technology relied on large manufacturers.

Motivated by this discovery, Chen decided to return to the United States as a civilian reporter. Using the contacts he had developed during his first visit, Chen connected with several companies in Boston's Route 128 and Silicon Valley. A particularly memorable visit took place at a magnetic superconductor manufacturer.* Chen was surprised to learn that the small company, which manufactured and sold equipment for nuclear fusion laboratories around the world, was run by a professor from Boston University. This was something that he had not seen in China, but that he discovered was a common practice in the United States. In China, culture and institutions prohibited professors from commercializing their research. In Route 128 and Silicon Valley, professor-entrepreneurs seemed to be the norm.

Chen could not stop thinking about what this meant for China. He started to see strong similarities between Boston's Route 128 and Beijing's ring roads. Route 128 is a peripheral highway that draws a semicircle around Boston. In the '70s and '80s it was densely populated by high-tech firms that had spun out from defense laboratories and local universities, such as Harvard, MIT, and BU.† Zhongguancun, the Beijing neighborhood where Chen's laboratory was located, also sits between ring roads in Beijing and is only a thirty-minute walk from two of China's top universities: Tsinghua and Peking.

Back in China, Chen could not stop talking about his trip to America. He insisted that the pool of talent they had in Zhongguancun was as good as the one he had seen in Route 128 and Silicon Valley. This was a bold statement. In 1980, the GDP per capita of China was an

* The Permanent Magnet Company.
† Boston University.

abysmal $430,* a small fraction of the $31,000 GDP per capita of the United States. But the physicist was convinced. He wanted to start a company and change China's research commercialization culture.

To gather support, he wrote a report for the China Association for Science and Technology and delivered it in a lecture on October 23, 1980. He observed in his report that in the United States, cutting-edge technology advanced quickly thanks to the rapid commercialization of research. This was driven directly by scientists and engineers who were eager to turn their work into commercial products. He remarked that the pool of talent in Zhongguancun was as good as that of Boston and San Francisco. But to compete with them, Chinese scholars needed to change their attitude and processes. During the lecture, Chen also made a surprising announcement: he was preparing to start a company in Zhongguancun.

But starting a company was easier said than done. Chen's original plan was to start a company at the Physics Institute. But his requests to do so were repeatedly ignored. Many of Chen's colleagues brushed off his ideas as an unrealizable fantasy. Luckily, he was soon able to find someone savvier than him.

Zhao Qiqiu led Beijing's Science and Technology Advisory Department. She believed in Chen's vision but also understood that starting a company was not the way to go. She advised Chen to avoid the use of the word 'company' and instead to set up a 'service department.' Chen took Zhao's advice and opened the Beijing Plasma Physics Association Service Department for Advanced Technological Development, which ran out of Chen's office and the Physics Institute's warehouse.

During the first few months, nobody made much of the Service Department. It was a bit of a hangout where staff were only allowed to work on evenings and weekends. But eventually, they started to make serious money. This didn't sit well with the people at the Physics Institute. In 1981, with activities extending from consultancy to manufacturing, the Service Department earned more than 30,000 yuan, at a

* Using World Bank estimates at constant 2015.

time when teaching was compensated at only 1.5 yuan per hour. This triggered a sequence of audits. Luckily for Chen, the audits brought Zhao Qiqiu back to the Service Department as a lead manager. Zhao tried to stop the attacks by negotiating a new deal. Chen's Service Department would avoid using any funds allocated to the nuclear fusion project and would record in its accounts the use of any instrument, even a pair of pliers, borrowed from the Physics Institute.

But Zhao's deal did not appease Chen's rivals. They demanded additional audits by the Physics Institute and the Chinese Academy of Sciences. The General Congress of the Physics Institute also began to smear him. They accused him of corroding the integrity of research by dishonestly acting as a middleman who was selling off decades of publicly funded knowledge.

The smear campaign worked. Chen's staff got scared and nobody wanted to walk with him around campus, even though many scholars in Zhongguancun were secretly following his story.

Among them was Wang Hongde, a computer electronics engineer who was only two years younger than Chen. Starting in 1979, Hongde had been training China's 're-educated' youth in the installation of computer rooms. But from a young age Hongde had been labeled a right-winger and counterrevolutionary and began experiencing a similar fate to Chen when his computer room installation operation began making a profit. This all goes to show that Chen's story was larger than him. His success would pave the way for others. His failure would bury their dreams.

By the end of 1982 the future looked bleak for Chen and his service department. But Zhao Qiqiu had one more ace up her sleeve. Her husband, Zhou Hongshu, was a reporter who had recently become deputy head of the Beijing branch of Xinhua News. One of the duties of this news agency was to prepare the 'Concise Report on National Trends,' a confidential document for senior members of the party. The January 6, 1983 edition of the confidential report included a story on Chen Chunxian. The story was written by Pan Shantang, a reporter who was part of Zhou's circle, and had been personally reviewed by Chen. Chen gave it the title 'Research Scientist Chen

Chunxian Sees the First Results of His Experiment in Disseminating Technology.' With this title, Chen struck the nail right on the head. In 1980, Chen Yun, a contemporary of Deng Xiaoping and one of the main strategists of China's famous market reforms, gave a speech emphasizing the importance of gradual change based on experimentation.[143] The speech was followed by a full endorsement from Deng Xiaoping. The 'concise report' was destined for an elite group of about 100 party members with the highest level of clearance. After reading the story, the vice-president of the state council, Fang Yi, invited Chen to his office for a meeting.* On January 8, Chen's efforts were commended by a member of the CPC's† central committee. By January 25, Chen Chunxian's story became national news.

The news came at the perfect time for Hongde, who only a month earlier had left the computing institute. In a speech to the Chinese Academy of Sciences, Hongde told the audience that he was leaving the next day and that the best thing for management to do was to agree with his formal departure. 'If that's no good, I will resign. And if you won't accept my resignation, then you should sack me.'[142] Hongde was determined to start his own computer-room engineering company. One year later, his company generated revenues of over 8 million yuan. His first contract, an order to rebuild Peking University's Honeywell computer system, was financed by a loan from the World Bank, which had made its first mission to China only a couple of years earlier.‡ By 1986, Hongde's company had reached a revenue of more than 50 million yuan. On the one hand, Chen Chunxian was a knowledge-rich entrepreneur who demanded better institutions. On the other hand, he was the little crack that broke the levee. A levee that was filled with knowledge.

*

---

* Chen had previously used a photocopied speech from Fang Yi in support of his ability to carry out scientific and technical consultancy work.
† Chinese Communist Party
‡ This was a mission led by Adrian Wood on October–November of 1980.[143]

The stories of Chen Chunxian and Wang Hongde include lessons about both knowledge and institutions. On the one hand, both Chen and Hongde were rich in knowledge. Chen had not only built a fusion reactor but was convinced that many of his colleagues in Zhongguancun were as capable as him. On the other hand, Chen and Hongde had to fight against institutions that were hostile to their entrepreneurial efforts. Even though they contributed to changing these institutions, their success was not the result of an environment that encouraged their entrepreneurship. On the contrary, they were ridiculed, ostracized, audited, and smeared. They were lucky to start their ventures at a time in which the higher spheres of the CPC had opened up to that type of modernization, and Chen was lucky to be able to reach them with his story. A few years earlier, the same initiative might have met a different fate.

Chen's story also highlights the importance of 'weak' institutions. One of the key contributors to his success was Zhao Qiqiu, an official who knew how to use and interpret rules 'creatively.' In a world of rigid institutions, Chen might have been unable to find the support he needed. This point is one of the main arguments made by Yuen Yuen Ang in her book *How China Escaped the Poverty Trap*.[144] There, the Singaporean political scientist argues that strong institutions exist to protect or preserve existing markets, but that you need weak or flexible institutions to create them.

The idea of flexible institutions is one that has gained momentum in the recent academic literature, but in China, it is an idea with a somewhat longer tradition. It is a notion that is integral to the *Guanzi* (管子), one of the foundational political and philosophical texts in China. While it is still a subject of debate, the *Guanzi* was probably written between the seventh and second century BCE. This 135,000-character long text has served – among other things – as a manual for the art of statecraft. In *How China Escaped Shock Therapy*,[143] Isabella Weber summarizes some of the key lessons of the *Guanzi*, which involve concepts that we know today in the West as the laws of supply and demand and the principle of monopoly rents.

The *Guanzi* doesn't talk about supply and demand, but about the balancing of prices using the 'light' and 'heavy' principle (轻重). Light is used as a metaphor to describe something that is unimportant, cheap, or trivial. Heavy means something that is valuable, important, or expensive. But the *Guanzi* doesn't describe products as universally light or heavy. The lightness and heaviness of products changes with the seasons and with market conditions. Grain is 'light' in the fall, during the harvest, but 'heavy' during the summer, when it is less abundant. Hoarding a product also makes it heavy. According to the *Guanzi* 'a ruler who is good at ruling the state must first of all enrich his people.' One way for the state to achieve this is to use the light and heavy principle, or 'qingzhong,' to flexibly balance commodity prices. For instance, the government must buy grain in the fall when it's 'light,' pushing its price up when the farmers need it the most. The *Guanzi* also says the government should sell its grain in the summer, pushing its price down when scarcity is at its highest. This is not a price fixation policy, but what many economists today would recognize as an anticyclical macroeconomic policy. The point of these examples is that the idea of balance and flexibility has a long institutional tradition in China. Zhao Qiqiu didn't need to learn this from Western scholars.

But are institutions the whole story? The point of Chen and Hongdu's story is not that institutions do not matter. On the contrary, they as struggle to change the attitudes of their colleagues and the general rules of the game played a pivotal role. But we must not forget that these are the stories of individuals operating at the cutting edge of technology. Chen and Hongde did not rebel to sell oranges at a traffic light. They rebelled to build companies that would supply equipment to plasma fusion laboratories and would set up state-of-the-art computing facilities. Even though they started as 'poor academics,' they were 'rich' in knowledge. Also, Chen and Hongde understood that, despite the cold shoulders of many of their colleagues, they were not alone in their struggle. These facts tell us that, just as we cannot engineer success by transferring institutions from a 'rich' to a 'poor' country, we cannot engineer success by asking developing countries to simply imitate China. The weak institutions that allowed a successful plasma physicist to succeed in Zhongguancun may lead to a very different result, in a place without the same concentration of talent.

This is an important point because the economic success of China has bred a generation of zealots who see Chinese institutions with the same glossy eyes that people saw Western institutions in the early 1990s. But development is about more than rules. The European reconstruction story has shown that finance is an important but not sufficient condition to engineer development. In postwar Europe, that financial aid was deployed in places that had a strong knowledge base, as illustrated in stories like that of Corradino D'Ascanio. These places also had a long tradition of functioning institutions. The key takeaway is that the financial aid received by Europe after the Second World War worked because these two other key pillars were already in place.

Institutions play an important role, but we should not expect them to work regardless of context. Mali and Chad are as poor today as China was in 1978, but my guess is that they do not have the equivalent of China's 1978 plasma physics program. To explore that idea more deeply, we need to think seriously about how to measure knowledge. How to capture whatever was present in a place like Zhongguancun. We need a way to understand the economic potential that was latent in China in the 1970s, and that might have been visible from activities such as those realized by Chen and Hongde. We need a formula to estimate the complexity of an economy.

# 9.

# A Change of Representation

Sometimes, the history of science is that of common words becoming formal concepts. Light, energy, and, more recently, information are colloquial words that scientists have gradually imbued with rigor and meaning.

Thinking about knowledge in quantitative terms may seem odd, but these etymological transformations are often slow. Consider the history of temperature, which took more than 1,000 years to become a number.

Today, we all think of temperature as something we can measure with great precision and can adjust by turning a dial. But for much of our history, our understanding of temperature was rather limited, and the word itself meant something quite different.

Back in ancient Greece, temperature was considered a quality not a quantity.* In his *Categories* and *Metaphysics*, Aristotle grouped temperature with other qualities, such as color and flavor. He did not think of cold as the absence of heat, but as a distinct and opposite force. Temperature was a mixture of these two invisible fluids. A mild temperature came from mixing 'hot' and 'cold,' just like gray paint comes from mixing black and white.

People in ancient Greece, however, did not use the word 'temperature,' since this is a Latin word. The word 'temperature' entered our vocabulary centuries later, when scholars translated the works of the Greek medical pioneer Galen around the twelfth century. In his work, Galen used the verb *krasis*, meaning 'to mix'. This was

* I base the beginning of this chapter on John McCaskey's excellent paper on the history of our understanding of temperature.[145]

later translated into Latin using the verb *temperare*, also meaning 'to mix', which was used to coin new nouns like *temperantia* and *temperatura*.

The idea of temperature as a mix of qualities survived for centuries. Prominent scientists such as Sir Francis Bacon and Robert Boyle subscribed to it.* Our understanding of temperature only really began to change during the Enlightenment, with the popularization of Daniel Fahrenheit's mercury-in-glass thermometers. This change accelerated with the work of Joseph Black, a Scottish physicist and chemist who helped uncover the difference between temperature and heat.†

Black figured out the zeroth law of thermodynamics, a principle that seems obvious, but which is key to understanding the difference between heat and temperature. The zeroth law is the idea that objects sitting together reach the same temperature. For instance, if you mix cold and hot water you get warm water. That is obvious enough. But what is not so obvious is determining the final temperature of the mix when this involves different materials.

In the eighteenth century, scholars like Black still thought of temperature as a thing. A material and invisible fluid that grabbed on to matter, like water grabs on to a sponge.‡ In Black's world, putting together two objects at different temperatures was like putting together a wet and a dry sponge. A hot object was a wet sponge soaked with the invisible 'temperature' fluid that made it hot. A cold object was a sponge that was rather dry. When you put them together, temperature flowed from the soaked sponge to the dry sponge until both of them reached the same temperature. That is a simple enough model. The genius of Black was in figuring out that you could use this idea to estimate the spongy capacity of any material by putting

---

* Galileo thought of temperature as part of a continuum (cold as the absence of heat), but according to McCaskey that was considered a minority view.[145]
† Black is also credited for the discovery of carbon dioxide.
‡ They had even conducted experiments 'proving' this idea by showing the ability of metal rods to conduct both heat and cold.

it together with others at different temperatures. For example, a kilo of hot water mixed with a kilo of cold water produces a result that is warmer than a kilo of cold water mixed with a kilo of hot iron.* This is because water is a better temperature 'sponge' than iron, and thus, holds more heat at the same temperature. Today, we call that sponginess the heat capacity of a material. In the case of water and iron, we know that water holds about ten times more heat than iron at the same temperature.

Separating heat from temperature was an important discovery, but also one that did not dispel the idea that temperature was a thing. You could still think of temperature as an invisible material fluid that grabbed on to water, iron, and everything else. But there were two important problems with this idea.

First, the fluid had to be massless, since temperature doesn't have a measurable weight. Second, people started to worry about the fact that you could extract an 'unlimited' amount of 'temperature' from materials with a rather small heat capacity. Consider making an iron cannon. In the eighteenth and nineteenth centuries, many cannon were made by boring a hole directly into what was otherwise a solid metal cylinder. As you can imagine, drilling a hole through a metal cylinder produces an enormous amount of heat that evaporates an enormous amount of water. But we know from Black's work that iron cannot hold that much heat. So where is all that heat coming from? Was it already in the metal? Or was it being produced by the act of boring? This observation helped luminaries like James Prescott Joule and Benjamin Johnson (Count Rumford) realize that heat was generated through friction. It was not a fluid that was in the metal, but something immaterial that was related to energy. By using ropes and pulleys to connect weights and water paddles, Joule estimated that a weight of about 900 lb falling 1 foot raised the temperature of 1 lb of water by about 1 degree Fahrenheit. This proved that temperature could be generated from mechanical

* Assuming that the kilogram of hot iron and the kilogram of hot water start at the same temperature (e.g. 90 degrees Celsius).

work, showing that temperature was not a 'thing,' even though all things have a temperature.

Going from there to a precise numerical definition of temperature still took a few more years. It was the life's work of William Thomson (Lord Kelvin) who spent much of his professional life looking for a scale with a better scientific foundation than the ones introduced by Daniel Fahrenheit and Anders Celsius (known at that time as the centigrade scale). Kelvin was looking for a scale where a change from 10 to 11 meant the same as a change from 100 to 101. His scale also needed to start at an absolute zero, meaning an absence of the quantity. This was an idea that Guillaume Amontons had introduced about 100 years earlier by using the known relationship between the temperature, volume, and pressure of air. Amontons calculated the temperature at which air pressure would vanish. That temperature turned out to be -273 degrees in the centigrade scale. With substantial help from Joule, Thomson put together many of these puzzle pieces.* And after several failed attempts, Thomson and Joule published a paper in 1854 comparing their temperature scale, based on an idealized Carnot engine, with the readings of a high-quality air thermometer. This led to an adjustment of the centigrade scale, which was redefined to match the scale introduce by Thomson and Joule and renamed to Celsius. This closed the more than 2,000-year journey that transformed temperature into a number.†

---

* Some of these pieces came from France, where Thomson worked at the laboratory of Victor Regnault. There, he learned about the work of Nicolas Sadi Carnot, a French military engineer and physicist who developed the theory of a maximally efficient heat engine. Carnot's theory provided Thomson a clean theoretical basis. Nicolas Sadi Carnot was inspired by his father, Lazare Carnot, a physicist and military leader who presided over the 41st French National Convention during the first republic period of the French Revolution. Working as a scientist, Lazare Carnot had estimated the maximum possible efficiency of a waterwheel. Heat engines were the waterwheels of the nineteenth century, so we can think of Nicolas Sadi Carnot's work as an extension of his father's work to the engines of his time.

† Of course, our understanding of temperature did not end there. Later in the nineteenth century, scholars such as Ludwig Boltzmann and Josiah Gibbs

But why should we care about the history of temperature? Isn't this a book about knowledge?

We should care about the history of temperature because it is the history of how we came to quantitatively understand something that is not a thing. Temperature is immaterial, but it is also not an invisible or massless fluid. It doesn't have a particle of its own, yet it is extremely prevalent. Even interstellar space – in its unimaginable emptiness – has a temperature that is above zero. The history of temperature, thus, represents an important paradigm shift. A natural philosophy embracing the fact that some 'things' are not things.

Temperature came first. Soon after, it gave us a way to think about information.* Information is also a thing that is not a thing, since it is not a fluid or a particle but a measure of how things are organized.† My argument is that knowledge is slowly moving into that camp. The camp where we have quantitative estimates of things that 'live' in things, as temperature lives in all objects or as information lives in a hard drive, but which are not things themselves. Knowledge lives in people and on the things we do. That again, is simple enough to understand. But putting a number on it is hard.

---

developed a theory from which you could deduce the ideal law of gases, and other key thermodynamic quantities, from first principles. In these theories, temperature is no longer thought of as a material fluid, but as a measure of how 'obedient' particles are to their energetic surroundings, which is related to how fast they are moving. Imagine a clerk stacking bowling bowls on shelves. In this analogy, a low-energy clerk represents a low-temperature system. That is a clerk who struggles with gravity and always puts the bowling balls on the bottom shelves. That clerk is obediently minimizing the potential energy of the bowling balls. A high-energy clerk represents a high-temperature system. That is, a clerk who not only puts bowling balls randomly on high and low shelves but constantly juggles them high into the air. Temperature is the excess energy of that nutty clerk.

* Since our quantitative measure of information is mathematically equivalent to Gibbs' entropy formula.

† My previous book *Why Information Grows* dedicates a few chapters to this idea.[84]

That's probably why our quantitative understanding of knowledge started with proxies, such as the measures of experience, cost, and performance explored in the learning curves of Thurstone, Wright, and Rapping. Formal work on relatedness, such as the product space, pushed us a bit closer to a more nuanced definition of knowledge. A quantitative estimate that is specific to places and activities. We can use measures of relatedness to estimate how close the economy of Paris is to the pharmaceutical industry compared to that of Bordeaux. If we had detailed data on China, we could use a measure of relatedness to see how far the computer industry was from Zhongguancun in the 1970s. But relatedness is also a proxy measure, based on the presence of other activities, such as plasma physics equipment manufacturing in the case of Zhongguancun.

But what we want is a measure that can help us estimate the value of knowledge agglomerations. A measure of the economic potential of Beijing, Tokyo, and Paris that simultaneously considers all of their activities. That is where we will find another historical connection, complexity shares a bit of etymological history with the word temperature.

When Gerard of Cremona translated Galen's work from Arabic instead of Greek, he used the word 'complexion' instead of temperature to describe Galen's idea of a mixture. The word 'complexity' may sound a bit mystical, but it is just a noun describing something composed of many interacting parts – like economies, which involve a deep division of labor and produce thousands of different products. Acknowledging that is acknowledging their complexity.

But since Gerard of Cremona, the meaning of the words 'temperature' and 'complexity' have somehow diverged. In fact, my argument in this book is based on the fact that temperature and complexity have gone in opposite directions. While both knowledge and temperature are 'not a thing,' knowledge, but not temperature, is the one I am equating to an infinite alphabet, since it is much more of a mixture than what temperature was once thought to be.

But where does the fragmentation of knowledge come from? Why is it made of multiple 'components'? Is the fragmentation of knowledge something fundamental? Or does it come from some inconvenient constraints? And could we – despite this fragmentation – quantify knowledge using a numerical scale? These are important but separate questions.

First, let's explore the idea of knowledge fragmentation.

Knowledge fragmentation emerges from the fact that the teams we form and the organizations we create can only hold a small fraction of all of the knowledge that is available.* Airbus knows a lot about air planes, but little about genetic engineering. Microsoft knows a lot about computers but little about furniture manufacturing. Like the proverbial sponges in our temperature example, individuals, teams, and organizations are limited in their capacity to accumulate knowledge. The key difference with the idea of heat, however, is that – unlike thermal energy – knowledge can be highly differentiated. People and teams do not simply accumulate knowledge but specialize in complementary flavors of it. That's why we see anesthesiologists working with surgeons and film directors working with actors.

Knowledge fragmentation doesn't play a big role in the theories of Thurstone, Wright, or Rapping, because these theories focus on the growth of the same 'flavor' of knowledge. But it becomes important when we consider knowledge diffusion across activities, which is the key point of departure for the idea of relatedness. Still, we need something beyond the idea of relatedness to estimate the value of the knowledge agglomerated in a country, city, or region. Something that starts from a nuanced representation like the networks used to estimate relatedness, but that flattens them into a single number. To get to those numbers, we will need one more lesson in physics. A lesson that goes beyond the physics of temperature and that can help

---

* Explaining the fragmentation of knowledge was one of the main points of *Why Information Grows*,[84] a book that explores the social and economic division of knowledge, and how institutions, such as a society's level of trust, limit our capacity to form the networks we need to accumulate vast amounts of knowledge.

us understand how seemingly difficult problems can become simpler once we discover their own internal alphabets.

★

In the early 2000s, I was loving my physics degree. An undergraduate degree in physics is a beautiful journey through the history of a discipline that has experienced many paradigm shifts. In 2001 I was preparing to learn quantum mechanics by taking a mandatory two-semester course on 'Methods for Mathematical Physics.'

I was lucky to have Rafael Benguria as my 'Methods' teacher. Rafael is a physicist with tremendous mathematical ability. To this day, I remember him as the person who guided us through the difficult journey that took us away from the physics of cannonball trajectories and into the math of 'spaces,' where you learn that each problem has its own 'alphabet.'

But what do I mean by an alphabet?

Maybe the easiest way to understand this concept is through music theory. Music theory starts from an alphabet that is related to the physics of vibrating strings. These are the musical notes, that in continental Europe we call *Do*, *Re*, and *Mi* and in the United States and England we describe using the letters *A* to *G*.★ With added sharps and flats, this is an alphabet of twelve letters. Twelve letters we can use to describe the sounds made by a taut string.

What we need to understand here, is that once you fix the tension and length of a string, it can only vibrate at a few well-defined frequencies. This might be counterintuitive, since you may think that you can jiggle a taut string at any frequency you want. But that's not how the world works. A taut string will only vibrate in some specific frequencies. Frequencies that define a metaphorical 'alphabet.' They are like Democritus' proverbial 'atoms,' but instead of being hard indivisible particles they are waves with unique lengths. Like all creative endeavors, music is infinite, but we construct it using

★ Do=C, Re=D, Mi=E, Fa=F, Sol=G, La=A, Ti=B.

*Figure 16: The fundamental mode of a wave, or first harmonic, is a wave with a single hump that vanishes on both ends. It technically uses only half of the string, so it has a wavelength that is twice the length of a string.*

an alphabet of twelve letters,* and we use twelve letters for a very good reason.†

Consider a guitar or piano string. These are taut strings tied at both ends. The first letter of that string's alphabet is the simplest wave that string can make: a vibration that moves it fully up and down, like in figure 16. That is called the 'fundamental mode' or 'first harmonic.' Its wave length is exactly twice that of the string, since to complete the wave you need to go up and then down. Since we can give this wave any name we want, we will call it a *Do* or *C*.

The next letter is a wave with two humps. This is the second harmonic. For a physicist, this is a new letter. But for a musician, it is a higher pitched version of the same note as before. A more shrilling *C*. The difference between these two *Cs* is what in music we call an 'octave,' a fundamental interval that repeats over and over as you move up or down the piano.‡ All of the twelve letters of our alphabet will fit inside this 'box.'

---

* This is technically true for instruments like the piano or the guitar. Instruments like the trombone can make continuous tones.

† Of course, there is more to music than the twelve notes of the chromatic scale, such as beat and rhythm, or the timbre of a particular instrument. Here I am focusing on notes or tones, which are basic building blocks for melodies and harmonies.

‡ In music, every time we double the number of 'humps' in a wave we are cutting its wavelength in half and doubling its frequency, resulting in a higher-pitched version of the same note. So, in this example our collection of *Cs* is represented by waves with one hump, two humps, four humps, eighth humps, and so on.

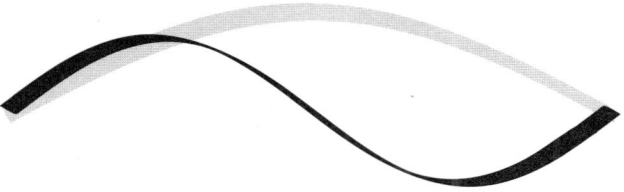

*Figure 17: The second harmonic (in black) is a wave with a length equal to that of the string.*

The next letter is a vibration with three humps. This sounds differently enough to count as a new letter in music. It is a wave that is two-thirds the length of the previous one. A very important interval called a 'perfect fifth.' If you've ever heard of major and minor chords,* these are built using a root note (a *C* in our example), a perfect fifth (a *G* in our example), and one more note, which is the one that makes it a major or a minor chord (respectively an *E* or an *E flat* in this example). The perfect-fifth or *G* is the second letter of our musical alphabet. But how do we go from two to twelve?

To get all twelve notes you simply need to keep on looking for perfect fifths. After twelve steps you will be back to the beginning. If we start from a *C*, we get to a *G*. If we start from a *G* we get to a *D*. If we start from a *D* we get to an *A*. And if we keep on going, we will eventually get to an *F*, which will bring us back to a *C*.† The physics of a vibrating string defines the finite twelve-letter 'alphabet' known as the chromatic scale.

The chromatic scale is a musical alphabet that we can trace back at least to Pythagoras. But the reason why it is an excellent point of

---

* In simple terms, chords are combinations of notes that sound good when played simultaneously, as when strumming a guitar or pressing a set of piano keys at the same time.
† The full circle, known as the circle of fifths, is: Do-Sol-Re-La-Mi-Ti-Fa#-Do#-Sol#-Re#-La#-Fa-Do or C-G-D-A-E-B-F#-C#-G#-D#-A#-F-C

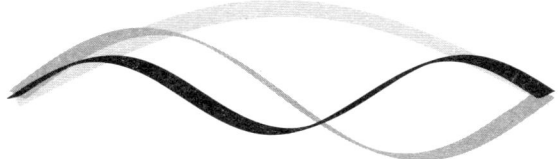

*Figure 18: The third harmonic (in black) is a wave of two-thirds the length of the string.*

departure for us is because it shows us how something seemingly complicated, like the shape of any vibrating string, can be decomposed into a natural alphabet.

But there is something that make these physics problems hard to solve. The waves I drew above are not just any waves, but those that satisfy two conditions. The first one is having a hump on either end of the string. This is the condition that makes this 'alphabet' discrete. But in principle, there are many shapes that satisfy this condition. We could have used a sawtooth wave with a pointy corner or a parabola for the first harmonic. Neither of these would work, however, because the waves need to also satisfy the equation describing the physics of the problem. This second condition is important because it rules out all other squiggly lines that vanish at the two ends of the string. Only the ones that satisfy both conditions count as an alphabet.

In Professor Benguria's course we learned many of these 'alphabets.' Consider changing a piano string for a drum. That change of geometry leads to a different equation and a new set of waves. These are no longer the 'sines' and 'cosines' that solve the wave equation for a string, but 'Bessel' functions (see figure on p. 133), the 'alphabet' of a drum. Every week Professor Benguria would teach us about one of these 'alphabets.'[*] Alphabets we needed to learn for our upcoming courses in quantum physics.

These functions may sound complicated, but the lessons we want to extract from them are simple. In fact, I just want

[*] Alphabets that in physics are known as 'special functions.'

| | Wave Length | Frequency Ratio | Musical Note | Musical Interval |
|---|---|---|---|---|
| | 2 | 1:1 | Do or C | Unison |
| | 1 | 2:1 | Do or C | Octave |
| | 2/3 | 3:2 | Sol or G | Perfect 5th |
| | 1/2 | 4 | Do or C | Two Octaves |
| | 2/5 | 5:4 | Mi or E | Major Third |
| | 1/3 | 3:1 | Sol or G | Perfect 5th |

Figure 19: Table of first harmonics.

you to focus on two things. The first one is the epistemological leap marking the transition from the classical mechanics of Isaac Newton to the quantum mechanics of Max Planck, Erwin Schrödinger, and Werner Heisenberg. Quantum mechanics, as is implied by its name, requires us to accept a discretization of the world. A world made of 'quanta,' or discrete units. Although we still use Democritus' word 'atom,' quantum mechanics is not about tiny spheres. It is based on discretizations that come from mechanisms just like the one I described for the musical alphabet. For example, the energy levels of an electron in an atom come from an exercise that is conceptually quite similar to that of finding the vibrations of a taut string. The equation you need to solve is more complicated, like when we moved from a string to a drum, but at the end of the day, you still get a collection of discrete 'waves' or 'levels.' In music, we called these discrete units C, G, and E. In a hydrogen atom we call these energy levels 1s, 2s, and 2p. In both cases, these are made-up names. What they really are, are discrete solutions to an equation. Discrete solutions that we can use to describe any vibration on a string or the energy level of any electron. These solutions are the 'alphabets' that correctly describe each problem.

In particle physics this idea is taken to the extreme. Not only the states that the particles occupy, but the particles themselves are solutions to even more complicated equations. More than 2,000 years after Democritus, scientists figured out that the world is not made of tiny indivisible balls, but of the solutions to equations summarizing the fundamental symmetries of nature.

The second lesson I want to rescue from these examples is the idea of 'mathematical representations.' This is the idea that to address problems mathematically we need to first find a good way to describe them. This sounds like a trivial point, but it is not. In fact, recent breakthroughs in science and technology, such as the creation of large language models, hinge largely on changes in representation.*

I'll assume that all readers are familiar with some large language models, or LLMs, such as ChatGPT, Grok, or Perplexity.

---

* Finding a good representation is not always easy, but it can help unlock a substantial level of understanding. All of the examples I have presented above are illustrations of good representations. Imagine trying to describe the shape of a vibrating drum using straight lines and $x$ and $y$ coordinates. Technically, it should be possible, but it would require pages and pages of algebra, as you try to fit every kink of the curvilinear drum surface with a line that has a different slope. Using Bessel functions and polar coordinates you can simply say that the shape of a drum is, for instance, one half of Bessel function zero plus one half of Bessel function two, and you are done. The correct representation can make a task that seems impossible trivial. That is why the idea of representation, and of learning representations, has become so important in computer science, as it is a linchpin concept for the development of artificial intelligence.

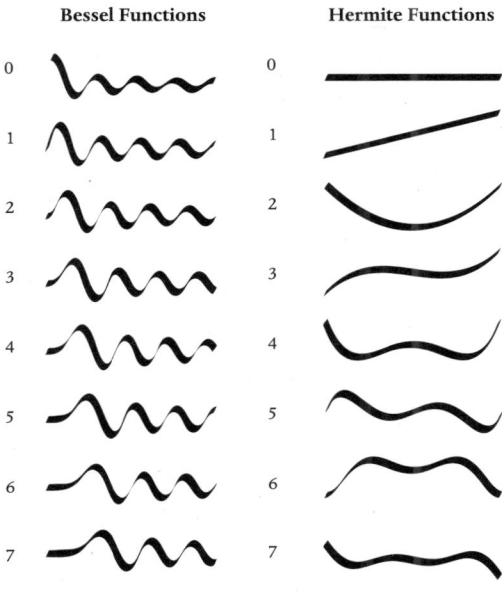

**Bessel Functions**  **Hermite Functions**

*Figure 20: Table of the first seven Bessel and Hermite functions.*

What readers might not know is how changes in the way we represent words paved the way for their development.*

The original attempts to represent words mathematically were based on naïve representations. These were vectors that were equal to one for a specific word and zero for all other words. For example, if we consider a text that uses 1,000 unique words, we could represent each of them with a vector that is 1,000 rows long. Each vector will have a 1 in the row representing that word and a in all other rows. In principle, this representation is good enough to numerically describe all of the words in a text. But in practice, it is not a good representation because the distance between every pair of words is exactly the same. The words 'coffee' and 'tea' are at the same distance than the words 'coffee' and 'diphtheria.'

A good representation should capture semantic relationships.

---

* Or parts of words known as 'tokens.'

Words are not independent but carry a meaning that is related to the 'company they keep.' Linguists back in the 1950s, such as John Rupert Firth, understood this. The practical application of Firth's idea, however, only emerged more recently, with the development of 'word embeddings.' These are numerical representations that force thousands of words into vectors that are only a few dozen rows deep. By forcing words into these reduced spaces, these vectors lump words that are used in a similar context, such as 'tea' and 'coffee,' or 'house' and 'home.'

The point is that representations matter. Changes of representation can be extremely important, and in the history of science they are related to paradigm shifts. By changing the way we describe the world we change what questions are easy or difficult to explore.

Today, word embeddings are a key step in the construction of large language models. But there is an important difference between these clever representations and the ones we constructed for taut strings.

In physics, we know that a representation is correct if it solves the right equation: the wave equation for our vibrating string or Schrödinger's equation in quantum mechanics. But we don't have the equivalent of Schrödinger's equation for language. So, scientists and mathematicians need to find representations that are good enough using a process that is more inductive. The good news is that this still works. Today we have representations that are able to capture meaningful relationships. Representations where 'Paris' is to 'France' what 'Madrid' is to 'Spain,' like a *G* is to a *C* what an *A* is to a *D*.

As someone who has been working on questions of economic development for the best part of twenty years, I can confidently say that, at least as of today, the idea of representations is not as central to economics as it is to physics and computer science.* For centuries, we have chiseled the stone of economics using the same

---

* In computer science there are entire conferences dedicated to learning representations.

representations, such as the ideas of labor and capital. But the transition from classical to modern physics involved a difficult leap of faith. A moment where scholars had to transcend the comfort of describing the world in terms of macroscopically intuitive concepts, such as solids, liquids, and gas, and embrace a world made of the solutions to 'exotic' functions. Biology had to undergo a similar transition when it accepted the bizarre nature of genes. These leaps of faith brought us to realities with exotic names, such as leptons and quarks in physics and BRCA-1 in biology. But if the economy is built on knowledge, and knowledge is as complex and fragmented as the words in a dictionary or the particles in a solid, we may need to consider representations that go beyond familiar quantities.

## 10.

# *The Knowledge Formula*

Two years had passed since Ricardo and I first met at the Kennedy School. We now had a relationship that was marked by the success of our first paper. On that particular day, we were looking for our next challenge. We wanted to do something together, but this time, Ricardo did not have a story about monkeys to tell.*

For a few years now I had been exploring economic data, so I started to complain about the fact that while there were plenty of indicators for countries, we had very few indicators to describe the products they made. That seemed like an opportunity to me, since as a young scientist I didn't just want to look under the proverbial lamppost. I wanted to light a new one.

Since we had been working with trade data, the simplest indicator we could construct for a country was its diversity: a count of the number of products a country was specialized in. But what was the equivalent for a product? I reasoned that the equivalent was the number of countries specialized in that product. For example, the number of countries specialized in the exports of

---

* Ricardo and I owe an important part of this work to Dani Rodrik. Rodrik is a Turkish-American economist who has been working for decades on questions of economic growth and development at the Harvard Kennedy School. Early in 2006 Rodrik released a paper asking: 'What's so special about China's exports?'[141] In that paper, Rodrik used an indicator he had developed together with Ricardo Hausmann to explore whether China's exports were typical for a country with its level of income.[146] The indicator, which they called EXPY, was simply the average income per capita of the countries exporting a similar basket of goods. In 2003, when China had an income per capita similar to that of Peru and Paraguay, its exports were as sophisticated as those of Spain and New Zealand.

cars, wine, or bananas. But since this was starting to become a bit of a tongue-twister, we needed a name for it, so I suggested 'ubiquity.'

The idea of ubiquity was simple enough, but I was standing in front of a whiteboard and wanted to make an $x$–$y$ chart for all products. So, I needed another variable for the $y$-axis. For countries, that was easy. You could compare a country's diversity with dozens of variables, like its population or GDP. For products, we had nothing. So, we started thinking about making a second measure to complete that chart. In hindsight, what I am going to tell you feels obvious, but it probably took us about thirty minutes to an hour to come up with a solution. Eventually, I settled for using diversity again. We could complete the chart by putting on the $y$-axis the average diversity of the countries specialized in a product. That was a bit more of a tongue-twister, but it led to the idea that we could do the same for countries, putting diversity on the $x$-axis and the average ubiquity of its products on the $y$. I drew the two charts with a worn-out marker and we reasoned about each of the four quadrants. Where would countries and products be located in these charts? We didn't have a clue. For that, we needed data. But that evening we imagined countries and products could be anywhere. A country far from both axes would be making many products that many others make. A country close to the origin would be making a few products that few others made. It was time to call it a day.

That night I went home to my apartment,* sat on my bed, opened my laptop, and loaded the data. After a few hours, I had made the calculations and the charts. To my surprise, countries and products were not spread across all four quadrants. Both charts looked a bit like a triangle, since countries that made many products tended to specialize in those made by few other countries. I pasted the charts into a PowerPoint presentation, sent it by email to Ricardo, and went to sleep.

Those first charts got the ball rolling. They were an interesting representation of countries, but we didn't really know what they

---

* Located near the intersection of Huntington Ave and the Riverway.

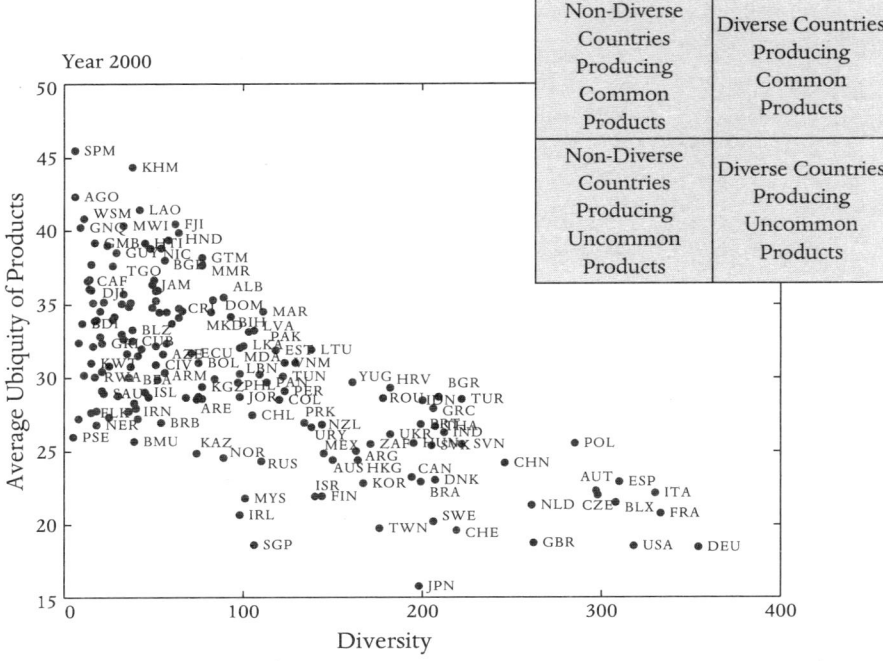

*Figure 21: Diagram comparing the average export diversity of each country with the average ubiquity of its exported products for the year 2000.*[147]

meant. Working on a new representation was interesting to me as a physicist. It also reminded me of our work on the product space, which also involved a new representation: the use of a network to estimate the affinity between a country and a product.

The problem here was that it became hard for us to continue moving forward using only words. The first variable for a country was its diversity. That was easy enough. Then we had the average ubiquity of the products it was specialized in. That was a bit harder to say. One step further resulted in the average diversity of the countries specializing in the same products.* Each step was getting harder to verbalize. We were getting stuck again.

One evening, I stopped at my friend José Miguel Fernandez's

---

* Weighted by the inverse of the ubiquity of each of these products.

apartment. José and I had been study buddies in Professor Ben-guria's methods class in Chile. We were now both in Cambridge, Massachusetts drinking beers and talking about our research.

A few minutes before midnight, I began walking to the subway. During that walk, on the corner of River and Auburn, I had the *aha!* moment we had been waiting for. We had to forget about language. Who cared about words such as 'diversity' or 'ubiquity' or things such as 'countries' and 'products.' At the end of the day, we were working with nodes in a network. Diversity and ubiquity were words we were using to talk about the number of links each node had, the number of products a country was specialized in, and the number of countries that specialized in a product. The next steps that we were struggling to verbalize were simply the average number of links of the first neighbors, second neighbors, and so forth. If we just thought about this in terms of networks, we could easily go to infinity and only worry about the name once we got there. It was a ludicrous way of thinking, but one that turned out to be correct. As soon as I arrived at my apartment I sat down again on my bed and worked on another presentation. The good news was that, as in the music example, infinity was not far away. After doing the same calcu-lation about twenty times the numbers no longer changed.

That 'infinity' representation, what today we call the Economic Complexity Index (ECI), turned out to be quite interesting. Sorting countries by diversity results in a relatively obvious ranking where 'big' countries like China, the United States, and Germany come on top. Diversity is a rough measure of size. But in the 'infinity' represen-tation, smaller yet sophisticated economies like Finland, Sweden, or Singapore bubbled up. At the same time, rich economies with a heavy dependence on natural resources, like Kuwait, Qatar, or the UAE, ranked much lower than their incomes. The infinity representation was capturing something that was different from size and income, but we didn't really know what it was.

One problem was that it was hard to communicate this number to economists. Unlike scholars in physics or computer science, whose interests are much more easily piqued by a new or unusual representation, the proverbial lamppost of mathematical representations in economics was not fully lit. Our colleagues were always trying to reduce our number to something they knew. The reactions we started to get was: 'Is it education?' 'Is it capital?' 'Is it financial market access?' 'Is it institutions?' And the answer to all of those questions was, well, 'No.' Our best intuition was that we were getting a measure that combined 'all of the above,' a composite measure of all of the complementary things that need to be present for an economy to succeed. So as a physicist trained in networks and complex systems, I started calling this infinity representation a measure of 'economic complexity.'

That summer I completed my PhD in physics and moved to the Harvard Kennedy School for a postdoc. I brought with me a complete draft of what would eventually become the 'economic complexity paper.' But we still needed one more figure. A figure showing that this new number could explain something that people cared about. Eventually, we found that our 'infinity' vector was a strong and robust predictor of a country's future income. This made it a measure of economic potential – of the ability of an economy to generate income that had not yet been realized. A measure of the potential of places like Zhongguancun.

I wrapped up the article by building a mathematical model and submitted it for publication. The study found a home at the *Proceedings of the National Academy of Sciences*, where it was handled by the Cambridge University economist Partha Dasgupta.[147]*

★

* Soon after the economic complexity paper was published, I went on one of the longest trips of my life. With a rollaboard, a backpack, and a Kindle filled with non-fiction books I traveled continuously throughout Europe for more than fifty days. Once I returned, I visited Ricardo at his office and told him that we should use our new measure to write a report. Why wait for it to be adopted by the UN, the IMF, or the World Bank if we could write a report ourselves? This eventually led to the creation of *The Atlas of Economic Complexity*, a report that we self-published with Puritan Press in 2011 and then published as a book with MIT Press in 2014.

Unlike the product space contribution, which was easier to understand, the idea of economic complexity was harder to grasp and thus was adopted more slowly. Not only was the concept more obscure, it was also a bit ahead of its time. Today, we know that this 'infinity' vector is what we call an embedding, like the ones used by generative language models to describe words. In our case, instead of looking at what words appear together in a sentence, we were looking at what products appear together in a country.*

But what set the index apart was its ability to explain future economic growth. Economic complexity seemed to capture a fundamental aspect of a country's economy, like the root note of a musical chord. A way to score the words that an economy makes when it plays Scrabble with the infinite alphabet.

This invited us and other researchers to explore new avenues of research. On the one hand, we became interested in whether measures of complexity explained other outcomes, such as income inequality or greenhouse gas emissions. On the other hand, we became interested whether we could derive complexity measures from other data sources, such as data on employment, patents, or publications.

* The idea of using embedding to represent words had been explored in computer science since at least the 1980s. But it only jumped into the limelight about five years after our paper, when in 2013 a team of researchers at Google introduced the now famous Word2vec.[148] That meant that for most of our colleagues, there was no reason to believe that our 'infinity' vector was anything more than a statistical curiosity.

Fifteen years have passed since we published our paper, and today, we know better. We understand that the Economic Complexity Index is the 'spectral embedding' of a matrix of similarities among countries. Thanks to the work of some colleagues in Vienna and Rome, we also know that ECI is the solution to a very fundamental problem.[149] Imagine you want to find a representation for each country and each product, so that the number you assign to each country and the numbers you assign to each of its products are as close as possible. This is a hard problem, because every time you change the number you assign to a product, you need to readjust the numbers you assign to each country. Yet, the solution to this problem is exactly the Economic Complexity Index. This means that the Economic Complexity Index is the best way to describe a country in terms of the products that it makes.

Connecting economic complexity with income inequality was an interesting question that we explored together with Dominik Hartmann, Cristian Jara-Figueroa, and other members of my team. Our research showed that economic complexity was a strong predictor of international differences in income inequality,[150] opening the question of: why?

This brings us back to Chen and Hongde's story. The idea is that knowledge-intense workers demand a different set of institutions. Chen and Hongde wanted meritocratic institutions that supported entrepreneurship. They wanted the economic freedom that Chen had seen on his trips to the United States. But that is on the side of the entrepreneurs. On the side of the workers, knowledge-intense industries also need to be more responsive to the demands of those with skills. In a peach-packing plant, the difference in productivity between the best and worst peach packer is tiny. In a software firm, the difference in productivity between the best and worst engineer is often large. Knowledge intensity must co-evolve with the meritocratic or inclusive institutions touted by Acemoglu, Johnson, and Robinson. So, if measures or economic complexity capture something about the knowledge implicit in an economy's activities, then they should also provide information about the institutions present in an economy. This co-evolution is well illustrated in another story from Polaroid.

Remember Polaroid's trailblazing founder Edwin Land? He had a famous habit of hiring women for technical jobs back when this was not a common practice.[64] The first person to see an instant photograph was Eudoxia Muller, a researcher at Polaroid working on the project that later become one of Polaroid's most iconic products, the SX-70 camera. Another of Land's famous women hires was Meroë Marston Morse, a key contributor to the development of instant photography. Land hired her in 1945, right out of college at a time when the company was tightening from 1,200 people to 240.[156,157] In only a few months, Meroë was running Polaroid's transition from sepia to black-and-white. Land was not required to hire these women by a formal law or by an internal policy. He discovered that inclusive institution himself as a savvy entrepreneur. He understood the value of tapping into an undersubscribed pool of talent, figuring out an institution that worked well for his knowledge-intense business.*

A second line of research explored the creation of measures of economic complexity using different datasets. This includes measures of complexity derived from the geography of patents,[107] industries,[164,165] and research publications.[166]† Again, the methods seemed to work. Economic complexity not only explained differences in economic growth among countries, but could explain differences in economic growth among a country's regions, such as states in Mexico or provinces in Spain.[164,169,170] This validated the notion that it was a measure of economic potential.[171-174] A great example of this is the work of the economic historian Giacomo Domini, who brought economic complexity to a historical context. Domini used data on the countries that participated in Paris's trade fairs during the nineteenth century to construct a historical estimate of their economic complexity. He used this estimate to show that economic complexity in the nineteenth century strongly and significantly predicted the economic growth of these countries during the next century.[173]

Research at different scales has also validated the idea that economic complexity was not just a measure of size. Using employment data from the US doesn't result in a ranking in which the largest American cities come out on top. Instead of getting New York, Los Angeles, Chicago, Dallas, and Houston, we get San Jose (Silicon Valley), San

---

* Our work on complexity and inequality inspired other scholars to explore similar ideas. Soon after our paper on economic complexity and inequality, people started to explore the connection between economic complexity and greenhouse gas emissions.[158-163] The idea was that a dollar of GDP generated from the exports of a high-complexity product – such as integrated circuits – requires less emissions than a dollar generated from exporting a lower complexity product – such as iron ore.

† Additional examples include college majors[167] and open-source software production.[168]

Francisco, Boston, Los Angeles, and Seattle. When using patents, we get San Jose, Austin, San Francisco, Boise, and Rochester. If you are not familiar with the last two, Boise, Idaho is the home of Micron, the only large American producer of computer memory, and Rochester, Minnesota is home to the Mayo Clinic, one of the leading life sciences research centers in the United States.

Eventually, a new synthesis begun to emerge.[166] Complexity was no longer a way to connect exports to economic growth, it was a more general feature that could be derived from a variety of specialization matrices and used to explain multiple outcomes. What was key to this synthesis was that the measures of complexity derived from different datasets tended to complement each other.[166] Measures of complexity derived from trade data have one important problem. They are biased in favor of economies that are integrated with sophisticated neighbors.[175] Mexico's trade complexity is high because it is integrated with the United States.[175] Slovakia and Germany are a similar story. But

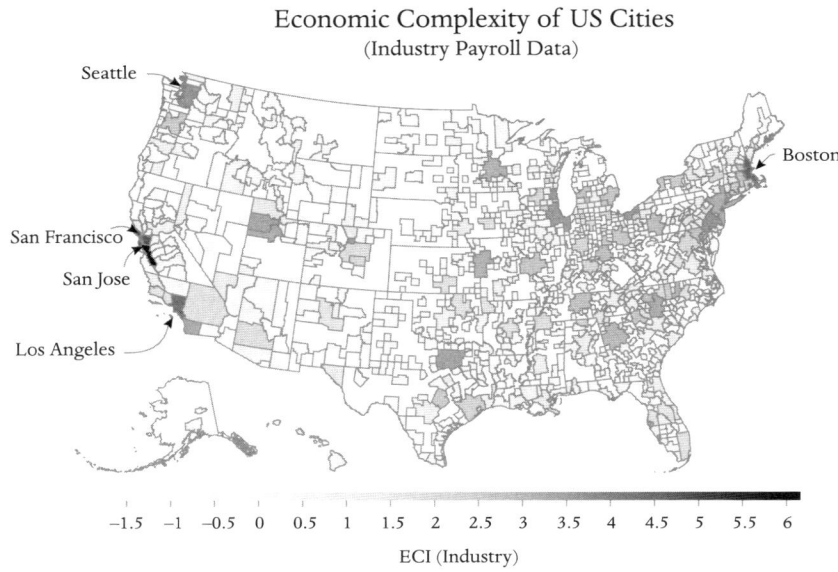

*Figure 22: Economic complexity of US cities (metropolitan statistical areas (MSAs) estimated using data on payroll by industry).[109]*

the biases of trade data are different from those affecting patents,* research publications, or software. So, when we put these measures together we get a picture that is better at explaining growth, emissions, and inequality than the one derived from trade data alone.[166,168]

While scholars were busy exploring extensions of these ideas, people in all types of began to adopt the concepts. Economic complexity became policy in manufacturing-heavy economies, such as those of Mexico and Malaysia,[176] and in resource-rich countries, such as Saudi Arabia and the UAE. Measures of complexity appeared to answer something that leaders in the developing world had long believed: the idea that knowledge matters, and that this knowledge is expressed in the activities that an economy is specialized in.

So, what does this formula tell us about China in the late 1970s? In 1978, the year that Chen visited the United States for the first time, China ranked forty-fourth in the world in terms of trade-based measures of economic complexity. It was about 0.3 standard deviations above the world average and not far from countries that were much richer than China at the time, such as Russia, Greece, and Costa Rica. Despite the crudeness of 1970s trade data, the knowledge formula seems to align with Chen's story. It shows that, even when the institutions were not in place, China had an untapped economic potential.

As this book was entering production, I dusted off some old calculations – an earlier attempt to develop a formal theory of economic complexity.†

Such a theory had long been an aspiration for those working in the field. The Economic Complexity Index (ECI) had become a widely used metric, but its theoretical foundations remained elusive. That was about to change.

The equations described a deceptively simple model in which a country's output depended on it possessing the capabilities required

---

* Patents, for instance, have the problem that they are not as prevalent as trade – which is something that all economies must engage in – and thus cannot be used to differentiate well among the lowest income / lowest complexity economies.
† These calculations had been done by Cristian Jara-Figueroa when he was my student, about a decade ago. We never submitted that work.

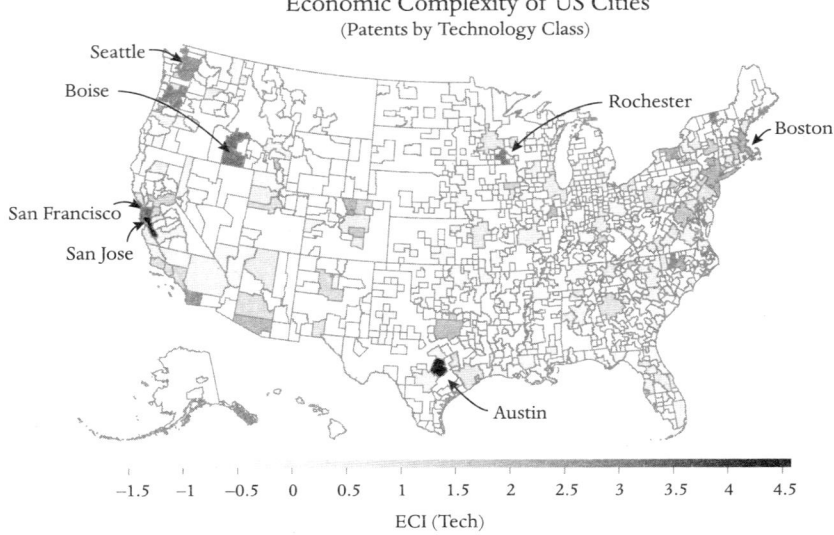

Figure 23: Economic complexity of US cities (metropolitan statistical areas (MSAs) estimated using data on patents by technology class).[109]

for each activity – much like needing the right letters to form words in a game of Scrabble. After adjusting a few key assumptions, I found I could estimate the mysterious eigenvector within this framework. That result established a clear mathematical connection between the theory and the metric we had been using in practice.

The math revealed that the index effectively distinguished economies with a high probability of possessing a given capability from those with a low probability. Extending the model to multiple capabilities, I realized the index was an estimate of the likelihood that an economy was endowed with many of them.

After nearly two decades, we finally had strong theoretical support for what the index had been measuring all along: economic complexity. A measure of all the above.

The mysterious eigenvector was no longer just a numerical artifact – it was implicitly counting the letters of the infinite alphabet, the many capabilities that make up an economy. It was like discovering the blueprint behind a fingerprint. The pattern had always been there, hidden in plain sight. Now, we had the theory to match the map.

## *The Third Law: The Principles of Value*

The third law tells us how to measure an economy's potential. In some ways, this is not a new idea. The idea that the potential of an economy is related to knowledge is something that we encountered early on in this book when we studied the work of Paul Romer. But Romer did not provide us with a method to estimate the knowledge base of an economy starting from the activities that it specializes in. To get that estimate, we had to take a detour into the world of mathematical representations and learn about the fact that many complex problems can be reduced to an alphabet. The representation we arrived at is one that predicts long-term economic growth and explains differences in key outcomes, such as inequality and emissions, validating the importance of this derived measure of economic potential.

But what does this measure really capture? We started this argument by talking about two historical approaches to international development: development through finance and institutional reform. Eventually, we came to accept that both approaches are incomplete, since the table of development needs at least three legs to stand. Knowledge, like that embodied in the story of Chen and Hongde, represents this third leg. Knowledge that co-evolves with institutions as shown in Polaroid's Meroë Marston Morse's story.

But are institutions and knowledge two different things? It depends on how narrowly or widely you define knowledge. If knowledge represents only the factual information that is available in scientific books, then institutions are clearly not knowledge. But if you take a wider view and define knowledge closer to the way in which some anthropologists define culture, as all the things humans transmit through non-genetic means, whether these are artifacts or rules, then institutions clearly fall into the bag.[177] Codified rules are certainly a form of explicit knowledge, and the way in which a judge, jury, or the general population interprets them are forms of tacit knowledge that are learned too.

The idea that a measure of economic potential based on

thousands of activities is a measure of 'all of the above' is still a powerful idea. It captures the agnosticism that helped physics and biology accept that the measures that count are not the ones that are easiest to interpret in terms of what we already know, but the ones that pragmatically explain the things we care about. If we give these measures a name, we will eventually get used to them. We will eventually learn a meaning, not by starting from a formal definition but by understanding how they work in practice. We can call musical notes whatever we want, but there is still beauty in knowing these are made-up names for the physical solutions to a vibrating string. We can call measures of complexity whatever we want, but there is still beauty in knowing that vectors extracted from matrices of similarity among cities, regions, and countries are robust measures of an economy's potential, capturing information that is hard to obtain through other means. There is something deep there.

This can be hard to accept because scholars like to have clear boundaries between concepts. But the universe doesn't care about our preference for clear boundaries. The world is often messier than what our theories allow us to see. It is useful to think of technology and institutions as two separate forces. But, not recognizing their co-evolution can be awfully naïve. That co-evolution, that intertwining, is where the potential of economies really resides. What Acemoglu and Johnson refer to as 'Power and Progress.'[178] That co-evolution is well-illustrated in one of our civilization's most important innovation: the history of religious institutions and the printing press.

*

Johannes Gutenberg is arguably one the most famous inventors in human history. That's why many people are surprised to learn that he died in poverty after losing his workshop to Johann Fust.*

* Printing provides a strong example of how innovation can shape institutions. This section follows Andrew Pettegree's masterful book on the story of printing: *The Book in the Renaissance.*[179]

Gutenberg was a genius. He didn't merely invent movable type – he reimagined the entire printing process. His multi-step casting technique produced type strong enough to withstand repeated use, and his quest for the perfect ink took a great deal of resolve. For years, Gutenberg honed his ideas while producing short texts, presumably in Strasbourg, until he returned to Mainz in 1448. There, he began working on a groundbreaking project: his eponymous Bible.

To make his Bible, Gutenberg needed financial help, so he teamed up with the bourgeois financier Johann Fust. Printing a Bible was a multi-year project that required building many presses and buying a copious amount of vellum, the animal skin used to print the luxury copies. It was a project that also required patience, since it was an investment that would only generate returns once the Bibles were completed. But the potential payday was huge. Gutenberg's Bible would eventually sell at a price of 20 gulden for the paper copies and 50 gulden for those in vellum, at a time where a stone-built house in Mainz was priced at about 80 to 100 gulden.*

Unfortunately, Gutenberg didn't get to enjoy the fruits of his labor. As soon as he finished his Bibles, Fust sued for the money he owed. The court rule in Fust's favor, sentencing Gutenberg to hand over his workshop to Fust.†

From that moment on, printing began to spread. In 1460, Albrecht Pfister was running what was probably the first printing press outside Mainz. Pfister may have gotten access to Gutenberg's original types after Fust and Schöffer discarded them. Printing also made it to Strasbourg around the same period, a city with connections to both Gutenberg and Fust.‡ A few years later, printing had made it to Cologne thanks to Ulrich Zell, an apprentice of Fust and Schöffer.

---

* The yearly salary of a skilled craftsmen was about 30 gulden.

† Fust would continue printing other works with Peter Schöffer, an apprentice of Gutenberg who later married Fust's only daughter. Schöffer's children would also work as printers. One of Schöffer's sons, Peter the Younger, was among the first to specialize in the printing of music.

‡ Fust was a member of the Strasbourg goldsmiths' guild.

The valley of the Rhine became a Renaissance 'Silicon Valley,' but the spread of this key technology soon escaped its riverbanks.*

The growth of printing was so explosive that book markets were saturated in a few decades. By the 1490s, books were piling up outside shops in the Piazza San Marco in Venice. By the early 1500s, Europe had reached a stable number of printers per capita.[180] In the language of the *Guanzi*, Gutenberg transformed something extremely heavy into something light. He had achieved a metaphorical alchemy that changed the balance of the European economy, even if he didn't get to enjoy his transmuted gold.

The impact of printing was massive. By the 1480s, it had increased the availability and reliability of astronomical data, paving the way for future discoveries by Tycho Brahe, Johannes Kepler, and Nicolaus Copernicus. By the 1500s, it was changing music through the development of printing types designed specifically for that purpose. But printing was only getting started. Cities that were geographically closer to Mainz, and thus adopted printing earlier, began to give birth to more famous artists and scientists.[180] The medium was changing the message, as Marshall McLuhan famously pointed out.[181] This eventually shook the foundations of European institutions, when in the 1510s Martin Luther really 'went viral.'

One of the most successful products during the early days of printing were indulgences – documents sold by the Catholic Church in exchange for forgiveness. Unlike books, which were major projects, indulgences were 'one-pagers' that were relatively easy to print. This made them attractive to both printers who were not experienced enough to undertake large projects and experienced printers who needed a cashflow infusion while working on books.

Indulgences were a huge business. Between 1498 and 1500, the Monastery of Montserrat in Catalonia commissioned more than 200,000 indulgences. In 1500, the Bishop of Cefalù in Sicily ordered 130,000. In

---

* The valley of the Rhine is still one of the most productive areas of Europe and the world, known colloquially as a part of Europe's 'Blue Banana.'

*The Book in the Renaissance*, Andrew Pettegree writes that 'of the 2,000 printed single-sheet items surviving from the fifteenth century over one third were letters of indulgence. Ninety percent' of them printed in Germany. So, when Luther published his *Ninety-five Theses* disputing the power and efficacy of indulgences, he was writing about something that was big business and was on everyone's mind.

Interestingly enough, Luther was not based in a big center of printing. He was based in Wittenberg, a small German town that had opened a press and a university in 1502. During that first decade, Wittenberg printers struggled to compete with more established centers of printing. That changed when Luther went viral. In 1517 he published his *Ninety-five Theses*, and in 1518 published more than eighteen original works, including his *Sermon on Indulgences and Grace*, which went through fourteen editions in its first year alone and eight more in the next. Major centers of printing in Germany, such as Leipzig, Nuremberg, Augsburg, and Basel, began reproducing Luther's work.* Data reported in Pettegree's book estimates that between six and seven million evangelical pamphlets were put into circulation during the first decade of the Reformation. That's viral, even by today standards. Luther became the most printed author of his time, not by a little, but by a lot. This meant that printing not only fueled one of the most important changes in the history of religious institutions, but also changed the world, by revealing the economic value of short texts.†

Communication technologies can influence institutions, since they not only affect how we communicate, but what we communicate. As technologies rewire society, they make us rethink the

---

* Pettegree also reports that between 1520 and 1525, Wittenberg's presses produced more than 600 editions, and Germany more than 7,764 editions, more than four times the number of books produced in Italy during the same period. Wittenberg produced at least 100,000 copies of the New Testament during Luther's lifetime.
† It is tempting to think of Luther's *Ninety-five Theses* as a long manuscript, but the truth is that most of his theses are a single sentence and all ninety-five of them fit on a few pages.

institutions that rule the world. It is hard to imagine the Reformation without Luther going viral. When we look at history long enough, we can see the arrow of causality moving in both directions: institutions enabling technology, and technologies shaping institutions.

But there are also simpler ways in which knowledge and technology can affect institutions. If institutions are the rules of the game, then we need to ask ourselves where people learn those rules. For most of us, that is either at home or at work. This means that institutional change often draws inspiration from practices that appear in the workplace. Chen learned about the institution of professor-entrepreneurs by visiting private sector companies, not the offices of an elected official. Land discovered the benefits of hiring women for knowledge-intense research jobs while working to save his own firm.

The tales of Bretton Woods and Zhongguancun serve as a reminder that while financial aid and institutional reform are crucial for economic development, they must be complemented by a rich reservoir of knowledge and a good deal of state capacity. We cannot assume that the former two factors alone will yield the same results in Beijing and Bamako. In the end, development is not a one-size-fits-all solution but a co-evolutionary process that requires a nuanced understanding of each country's unique context and capabilities.

Still, there is one important reason why we should focus on institutions. They might not be everything, but they are one of the few things we can try to change. This brings us to one of the last legs of our journey, a set of chapters exploring ideas on how to stimulate knowledge development while considering the principles that govern the growth and diffusion of knowledge.

# PART IV

## The Seeds of Knowledge

II.

# *Yachay Again*

If you could build Yachay again, what would you do differently? We can certainly use the laws of knowledge to understand what went wrong. These principles tell us that knowledge grows more easily under certain conditions. There is, so to speak, a direction to the metaphorical wind given by the learning curves of Thurstone or the knowledge diffusion stories of Slater and D'Ascanio. Yachay was a well-intentioned effort. But in a world where every economy is competing to attract or generate knowledge, intentions must be met with a sound strategy. That means building an innovation policy that considers the principles that govern the growth and diffusion of knowledge.

It should be clear by now that Ecuador's idea of building a city in a remote location was farfetched. If you've visited Quito or Guayaquil, you know that these are large cities with functioning commercial and residential real-estate markets. The decision not to bet on your two main cities was a key mistake. Ecuador's innovation ecosystem was not limited by a lack of buildings but by a lack of specialized human capital. So, if we had a second chance to invest a billion dollars in a program designed to kickstart Ecuador's knowledge economy, what could we do differently?

While Ecuador was busy planning Yachay, China was doubling down on Zhongguancun. In 2010, Zhongguancun was no longer a backwater of innovation. It had lived a defining moment in 2004 when Lenovo, a company that also had 'spun out' from the Chinese Academy of Sciences in the 1980s, acquired the personal computer business of IBM. This was by no means an easy feat. For decades Liu Chuanzhi used his charm and business acumen to keep Lenovo afloat in what was a very difficult environment.* But during the 2010s China supercharged the Zhongguancun innovation using clever policy interventions.

In *AI Superpowers*,[182] Kai-Fu Lee – a computer scientist and venture capitalist – tells the story of Guo Hong, 'a startup founder trapped in the body of a government official.' In 2010, Hong wanted to transform a 'slice of Beijing into the Silicon Valley of China.' This was a similar ambition to that of Yachay, but one that unfolded very differently. Both Guo and Lee agreed that the intense competition experienced by Chinese internet entrepreneurs during the 2000s had forged a new generation of business leaders. As a venture capitalist, Lee wanted to fund some of these entrepreneurs. Guo was instead looking for a reason to relocate some of these companies to Zhongguancun.

What Guo and Lee didn't do was build a city of knowledge in a remote corner of the Altay Mountains. Instead, Guo reconditioned a street in Zhongguancun with buildings designed to house startups, venture capitalists, and incubators. The Chuangye Dajie – the Avenue of the Entrepreneurs – was not far from Peking University, Tsinghua University, and a campus of the Chinese Academy of Sciences. The deal that Guo offered Lee was that companies moving into the Avenue of Entrepreneurs would get a discount on rent. Lee liked this idea, since many of the companies he was funding were spending money on rent that they could put to better use. This resulted in an incentive to concentrate venture capitalists and startups.

But the differences between Zhongguancun and Yachay went far beyond their choice of location. Chinese policymakers wanted to transform China from an economy in which growth was driven by manufacturing into one in which growth was driven by innovation.†

---

* The story of the early days of Lenovo and the adventures and misadventures of Liu Chuanzhi are well summarized in part four of Ning Ken's book, *Zhong Guan Village*.[142]

† That push resulted in the creation of more than 6,600 incubators during that period.[182]

That involved the creation of 'guiding funds' combining contributions from private investors and the government. Government contributions were introduced as a means to increase the potential upside of a fund without fully removing its risk. If the companies supported by the fund failed, everyone lost their money, including the government. But when the companies succeeded, the other investors could buy the government shares at a predetermined price. The result was the creation of funds in which the government got its money back, with a profit, and that incentivized private sector investors to join funds nudged by the government into certain sectors. This led to an explosion of Chinese venture capital. During the first half of that decade, venture capital in China grew from around US $3 billion to over 26 billion, as large cities and provincial governments were incentivized to copy Guo's idea and focus on 'mass entrepreneurship and mass innovation.'

Zhongguancun's story is also very different from that of Yachay in terms of its choice of sectors. In 2010, Lee and Guo focused on entrepreneurs working on smartphone applications. This was a smart choice because of demand and supply considerations. On the one hand, Lee and many of the entrepreneurs working in Zhongguancun were experienced in digital technology. On the other hand, the number of smartphones in 2010 was a small fraction of the number of smartphones that would be in circulation by the end of the decade. They were looking at a window of opportunity that was destined for growth.

The idea of windows of opportunity is one that has been explored by in depth scholars such as the Korean economist Keun Lee. Keun has long studied the process of leapfrogging.[183] This involves jumping a technological generation in an attempt to start early on the next learning curve. It is a risky strategy that Korea famously executed to overtake Japan in the production of LCD monitors.

Now, while the idea of leapfrogging technological generations can sound quite general, there are important strategic considerations that policymakers should consider. One of them is the frequency or length of their disruption cycle. LCD manufacturing is an example of a quick-cycle technology. Between 1990 and 2010, LCD manufacturing went through ten generations. Other technologies, such as those used to produce pharmaceuticals, are long-cycle technologies. In *A Shot to Save the World*,[184] Gregory Zuckerman provides a thorough account of RNA therapeutics, a scientific and technological

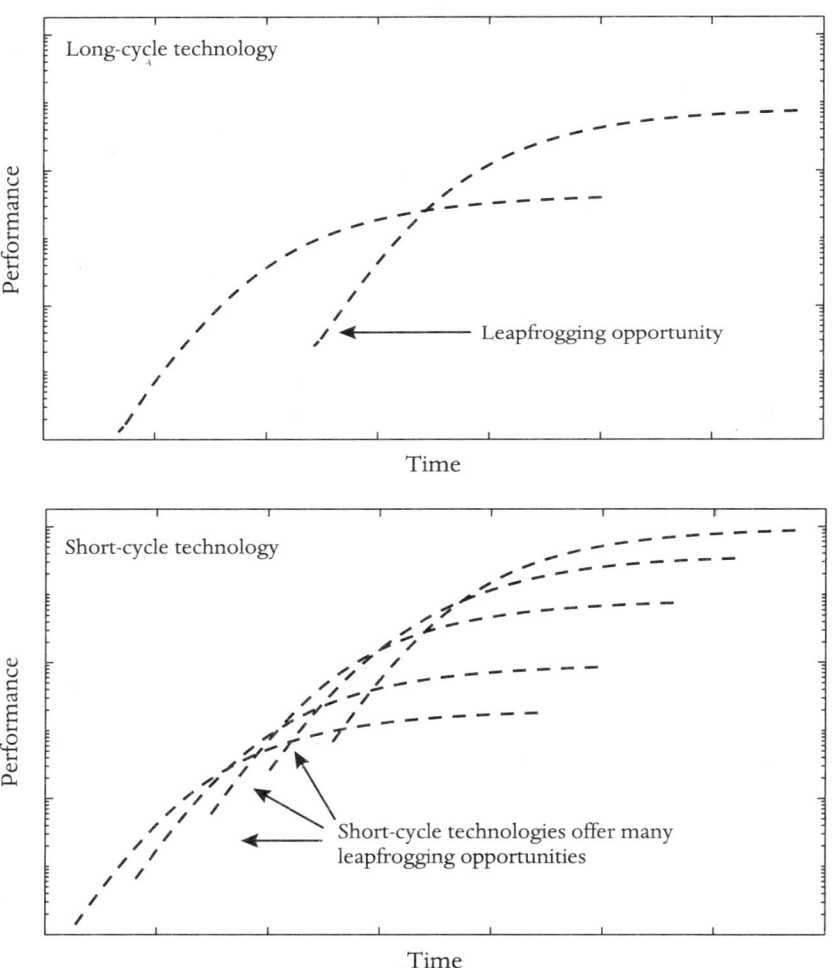

*Figure 24: Leapfrogging opportunities emerging in the presence of overlapping generations of technology. These opportunities are more frequent for short-cycle technologies.*

breakthrough that was decades in the making and that was instrumental for the development of the COVID vaccine.

Leapfrogging is easier when it targets short-cycle technologies, since these are less reliant on long-established industrial processes or infrastructure. Kai-Fu Lee's focus on smartphone applications coincides with this idea.[183] It was a chance to leapfrog into the future by focusing on a short-cycle technology.

The contrast between Zhongguancun and Yachay shows how efforts to 'engineer' a new Silicon Valley can differ enormously in terms of implementation. Zhongguancun's strategy follows what we would recommend based on the principles that govern the growth and diffusion of knowledge. It was an effort focused on an established innovation district leveraging related expertise to target a new window of opportunity. Certainly, Ecuador is not comparable to China.* But still, the Andean nation could have doubled down on Quito and Guayaquil and invested in guiding funds similar to the ones deployed in Zhongguancun. This second point is crucial, since Yachay's implementation included little investment and plenty of spending. Ecuador spent hundreds of millions of dollars on construction and salaries without a clear idea of how these 'investments' could generate a return. This makes the second part of the Zhongguancun story as important as the first. Supporting innovation is not just about finding the right geographical location but about figuring out a strategy to invest rather than spend the funds. This brings us to other things that Ecuador could have done differently.

Let's start by remembering that Yachay was an effort to grow knowledge Ecuador didn't have. This goes against the principle of relatedness, but it is still a valid and quite common policy objective. Most developing countries are in a situation in which they need to steer the knowledge base of their economies into unrelated activities. This policy objective is still compatible with the principle of relatedness, since this principle does not mean countries should only

---

* Ecuador's population of around 18 million is less than the population of the metropolitan area of Beijing.

focus on the activities that are most related to their economies. The only thing that the principle of relatedness does is provide a model for the headwinds or tailwinds involved in each investment. It is like a measure of the strength of gravity for someone looking to build a rocket. The good news is that the principles we studied earlier also have something to say about factors that contribute to the growth of unrelated knowledge. We know that unrelated knowledge often requires attracting people, as in Samuel Slater's story. That means that if Ecuador wanted to invest US $1 billion of taxpayer's money in an effort to become an international research hub, they should have focused on attracting international talent. But how?

To figure that out, we can start by looking at the global market for international research talent. Even though researchers concentrate in space,[16] they are one of the most globally mobile populations on the planet. So, our first clue is to figure out what is an internationally good deal for a researcher. In this example, I'll focus on academia, since this was one of the key targets of Yachay.*

Let's start our exploration in Europe. Europe has a very clear system of research grants with well-defined 'prizes.' The top grants in Europe are the European Research Council grants, or ERCs. These are extremely competitive bottom-up grants, that scholars from all academic disciplines compete for ('bottom up' means that scholars are not required to focus on any particular field or topic). That makes ERC applications a battle royale between particle physicists, political scientists, biochemists, and everyone else.†

In 2024 the funding ceiling for an ERC grant was about €2.5 to €3 million for a five-year award, or about half a million a year. As with most grants, these are not funds that go directly to the researcher, but that are used to support a research team.

---

* Although similar ideas could be used to design a program targeting private sector entrepreneurs.
† ERC grants exist at three levels – junior, intermediate, and advanced – depending on the number of years that have passed since a scholar completed their PhD. ERCs also include a competition for interdisciplinary teams, or synergy grants.

Another example is ETH Zurich, a prestigious research university in Switzerland. ETH provides chaired professorships with about a million Swiss francs a year.* In the United States, an RO1, a highly coveted career-making grant in the life sciences is similar in size to an ERC. So, you get the picture. For many research teams, a million dollars a year goes a long way.†

So now that we understand the international prices we can explore better ways for Yachay to invest a billion dollars. The first thing you want to do is to avoid spending it. Instead of putting that billion into new buildings, you can put it in an endowment managed by a professional fund. Let's imagine a fund with a payout rule of about 5 percent. This is a rather conservative estimate considering that the SP500 more than doubled in the decade that Yachay was built. The 5 percent rule means that the authorities running this innovation program would have a perennial ability to invest US $50 million a year. That means that after twenty years they would have invested the original billion and would still have a billion – or maybe more – left in the bank. Fifty million dollars may not sound like much, but it is about half of what Chile – a country with an economy that is about three times that of Ecuador's – spends every year in 'advanced human capital,' which means scholarships for PhDs, master students, and postdocs. So, in the context of Ecuador's innovation system, 50 million a year is quite a lot.

But how exactly would you disburse that 50 million?

If your goal is to attract world-class talent that can bring new knowledge, then you can use the endowment payout to create long-term international attractivity chairs. Here, I'll be a bit loose with the numbers, since what I am trying to do is get a general point across. Let's say we use the 50 million to create fifty chairs that are similar to those provided by ETH. The scholars selected for these chairs would be invited to set up their laboratories in Quito or

---

* About $1.1 million.

† Those funds are not destined for the professor or principal investigator, but primarily to finance a team composed of PhD students and postdoctoral fellows.

Guayaquil and would use their million dollars a year to rent office space (maybe at a local university), hire administrative support, and fund the salaries of their PhD students and postdocs. Of course, a plan like this would encounter many problems, from the selection of the principal investigators to their smooth relocation. But these are problem that we can deal with.

The first challenge requires both a strategy to find top researchers, and implicitly in that, a choice of research sectors. We can think of hiring a top researcher as a process similar to hiring top talent in business or sports. That means that you may want to avoid an open contest, as this would generate adverse selection – the most established PIs would not bother to apply.* Think about FC Barcelona or the Boston Celtics looking to hire a new star player. There is a reason they scout for talent instead of running an open tryout. Top talent should have a track record of productivity that should be identifiable by a good scout.† The second part of this process is the implicit choice of sectors. Following the 'short-cycle' idea means looking for researchers who would not require big investments in facilities or equipment. For example, a million dollars can go much further in computer science or computational biology than in genetics or immunology. In the latter, the need for specialized facilities and the constant need for reagents and animal models can put a lot of pressure on a lab's budget.‡

But finding a candidate is only one of many problems. Moving to

---

*An established PI would say: Why bother writing an application for a chance to get funds that will require you to move to a new country? The supply constraint is on the other side.

† In academia, a researcher's track record is well documented in their list of publications and their scientific impact (usually measured in terms of citations).

‡ There is also the need to manage the expectations of local researchers, who may react negatively if they think the project is running on funds they should get. This is related to the challenge of connecting the incoming talent to the local research ecosystem. One way to partly align some of those incentives is to provide local universities with some of these attractivity chairs. We can think of this as a 'coupon' that universities can strategically redeem once they identify a suitable candidate.

a new country will not be easy for international researchers who are unlikely to speak the local language and will be unfamiliar with the local bureaucracy. In the case of Ecuador, sharing a language with many countries is an asset they can leverage by narrowing down their search to internationally validated Spanish-speaking scholars. Still, the incoming scholars will need to get a driver's license, find a home, find a school for their children, open a local bank account, etc.* This could be alleviated in part by using some of the support to hire a local executive assistant dedicated to onboarding the incoming research team.† It could also be used as an opportunity to identify poorly designed bureaucratic procedures

But there are also long-term problems. The low transferability of pensions among countries constrains the mobility of people who have worked for a long time in the same place. For a researcher moving internationally, this could mean losing or diluting decades of pension-fund contributions.‡

Still, despite all of these problems, a project focused on attracting talent using the interest payments of an endowment would have better longer-term viability than a project focused on building a city of knowledge that is burning a limited pile of cash. Attracting and integrating world-class talent is hard. It requires thinking creatively

---

* When I moved to France in 2020, I learned that opening a bank account requires a local address, but renting an apartment requires having a local bank account. If you grow up in France you can start with your parent's address, but when you are totally new to the country you are stuck in a loop until someone decides to look the other way. These catch-22s are all too common for those who relocate, are often invisible to the locals, and are low on the political agenda of a country.
† Also, a special law facilitating the easy validation and transfer of documents – such as the driver's license – could go a long way.
‡ A common assumption of pension systems is that a worker contributes to the system in the same country throughout their lifetime. While there are some means to claim funds between agencies (e.g. among countries in Europe), people who have worked in multiple countries struggle to reconcile pension-fund contributions in different places.

about how to solve an ever-growing stack of problems.* But smaller countries such as Ecuador, which lack the population advantage of a place like China, have to figure out how to do exactly that. Still, the beauty of the idea is that, if managed responsibly, a ten-year trial of a program like this one would not cost much. Or at least, not as much as Yachay did. Even if it fails, at the end of the ten years, you still have a billion dollars!

*

In his 1998 Pulitzer Prize winning book, *Guns, Germs, and Steel*,[185] Jared Diamond borrows a line from Leo Tolstoy's *Anna Karenina* to explain why animal domestication has been historically difficult. Tolstoy's opening line, which depending on the translation goes something like this: 'Happy families are all alike; every unhappy family is unhappy in its own way.'

This is a powerful opening statement that Diamond uses to explain that all domestic animals share some key characteristics. They are like the happy families in Tolstoy's dictum. The flip side of this is that there are many ways for families to become unhappy, and many reasons why animals may be hard to domesticate. Some species are unsuitable for domestication because they lack an amenable social structure, like a herd, or because they are bad at reproducing in captivity. The general point is that, when success requires simultaneously meeting multiple criteria, failure can result from missing only one of them. Here, I would like to borrow Diamond and Tolstoy's idea to argue

---

* Some of these barriers can be deeply cultural. For example, in France, a common point of complaint among foreigners is the fact that people are bad at responding to emails. France's communication culture runs primarily by calling people on the phone or visiting them in person at their office. This can be frustrating for foreigners who are not fluent in French and need to rely on AI translation technology. Also, people who travel frequently and cannot easily make or receive calls or visit people in their office need to communicate asynchronously. The lack of email culture also makes it difficult for French research organizations to work with organizations outside of France, which expect to communicate via email and in English.

that economic development is somewhat similar, since all countries are developing countries in their own way. What I am trying to say is that the factors limiting the growth of knowledge can be highly idiosyncratic. So general rules, ideas, and programs, like the ones we have looked at so far, will not always take us far enough. At some point, we need to accept that idiosyncratic problems sometimes need idiosyncratic solutions requiring intuition based on a deep understanding of local conditions. This brings us to the importance of leadership in a world where not everything can be reduced to rules.

I recently learned about an interesting way to think about leadership from Paul Seabright, one of my colleagues at the Toulouse School of Economics. At one of our regular seminars, Paul presented some ideas he was working on for a book. One of these was a distinction between 'skill' and 'judgment.'[186] Skill, he said, is easy to judge, as you do not need to be a skilled juggler or basketball player to recognize someone with those skills. Judgment, however, is harder to identify, since it involves making decisions that people with poor judgment will not understand. Using sports as an analogy, you can think of skill as what the players have and judgment as what the coach provides. Certainly, one could think of judgment as a type of skill, but that is a semantic peeve that muddles the argument. Embracing the distinctions is more fruitful. It gives us a model of leadership as the formation and survival of networks built around people with excellent judgment. The quintessential decision-makers.

Putting these ideas together means that economic development benefits from leadership with good judgment. This may seem obvious, but there is an important nuance that comes from this argument, which is that copying the actions of a judicious leader – adopting a set of 'best practices' – is not the same as having good judgment. In fact, 'best practices' can grossly backfire when they are applied in the wrong context. Good judgment cannot be reduced to recipes. The judiciousness of the leader, not their actions, is what is hard to copy, and is what underlies their success.

A famous example of a creative and judicious leader is Lee Kuan Yew, who ruled Singapore from 1959 to 1990 and is considered by many to be the architect of Singapore's modern-day success. Singapore is a particular country. It has an integration policy designed to avoid ethnic enclaves by limiting the number of households of each race in a neighborhood or block. But there are many other famous judgment calls involving Lee Kuan Yew. Some of them might even seem strange to outsiders. For instance, when he was asked about things other than the multi-ethnic integration tolerance policy that contributed to Singapore's success, his answer was air-conditioning.[187] 'The first thing I did upon becoming prime minister was to install air conditioners in buildings where the civil service worked. This was key to public efficiency.' His idea was that 'without air conditioning you can work only in the cool early-morning hours or at dusk.' In his view, air-conditioning was key for the efficiency of the public sector.

Lee Kuan Yew was also involved in the construction of Singapore's award-winning Changi Airport.[188] After visiting Boston, Lee was inspired by the Hub's seaside airport. He realized that building a runway on reclaimed land had many benefits. It kept the noise of the planes away from the city and avoided the need for building height restrictions. Yew changed his mind on a dime at a time when Singapore had already started expanding its Paya Airport (today a military base).

The point is that Lee Kuan Yew's focus on air-conditioning and the construction of a coastal airport may be good judgment calls that may not translate well to other contexts. They are actions that were specific to Singapore's limited land area and tropical climate. Air-conditioning may have contributed to the growth of knowledge in Singapore, but it would probably be an ineffective policy action to take in many other parts of the world. As Tolstoy and Diamond remind us, every developing country is a developing country in its own way.

This means that promoting the growth of knowledge requires leaders with a good intuitive understanding of when the shoe fits. There are so many things that can potentially go wrong. I've experienced countries with idiosyncratic problems in their work culture, with many people being unable to communicate effectively and clearly through email. I've lived in countries that are unable to build public transportation systems. I've seen places with an excessive pride in their local language to the point where they struggle to make international business relationships. I've lived in societies that are sinking due to an overly cautious approach to digital technologies. I've worked with cultures in which lying is prevalent and expected. I've seen a little bit of everything by now.* Enough that I prefer not to name names, because my intention is not to point the finger at any particular country or culture. The point is that, at the end of the day, the actions a place might need to grow its knowledge base can be highly specific. Specific enough to require institutions or leadership with good enough judgment to make those. Countries can certainly borrow ideas from others, such as setting an endowment fund for the sciences or creating investment funds like those used by China to push its innovation economy. But as the saying goes, the devil is in the details. Ideas by themselves are no guarantee of results when their success depends critically on the quality of the implementation. If the available institutions are unable to understand, judge, and act on details, then the best ideas can become nothing more than good intentions whispered into the wind.

While broad principles, like the laws of knowledge, provide a vital framework, they need to be complemented by a nuanced understanding of local context.† Yachay's monumental infra-

---

* In a famous interview with Charlie Rose, Lee Kuan Yew shared stories of actions they took to help change key behaviors in Singapore's population, such as public urination.[189] Since some people were urinating in elevators, they installed devices that, after detecting urine, would stop the elevators, catching the perpetrators. Once perpetrators begin urinating outside elevators, they installed cameras as well. This is another example of the importance of state capacity, and in this case, creative enforcement, which is very different from on-paper regulations.

† Like when Clay Christensen told his story of steel mills to Andrew Grove from Intel, and allowed Grove to extract the lessons and define the strategy.

structure and poorly strategized spending illustrate the perils of a grandiose plan aimed at sidestepping local context by betting on a foundational utopia. Had Yachay focused on the nuances limiting the ability of Ecuador to attract world-class talent, it might have produced a more enduring legacy.

Ultimately, growing the knowledge base of an economy requires institutions that can identify the critical, context-specific bottlenecks that impede progress and have the judgment to know when to act in ways that may not be immediately understood or appreciated. For Ecuador, and other countries facing similar challenges, the future lies not in attempting to recreate the success of others, but in crafting a unique path rooted in the reality of their environment. A path that even some people in Ecuador have been able to blaze.

*

In 1972, the Ecuadorian physicist Santiago Gangotena had a dream. Back then, he was a young graduate student at the University of North Carolina's nuclear physics program. But his dream was not to build a nuclear power plant, but a flagship university back in Ecuador.*

When he returned home with a PhD in 1976, he brought with him an unrelenting resolve. He put his plan in place by creating the CPU Foundation, a Corporation for the Promotion of Universities that was an intentional pun on the idea of a computer's central processing unit.† The goal of the CPU foundation was to create a private university in a country that had been unwelcoming of competition in higher education. With CPU, Gangotena designed a liberal art-education curriculum similar to the one he had seen in the United States.

Creating a university in Ecuador in the early 1980s required congressional approval. So, in 1981, Gangotena prepared the extensive

---

* This section is built on the history of the University of San Francisco de Quito as written by its founder on the university's website.[190]
† In 1980.

list of documents required to submit his project to Congress. Unfortunately, his proposal was rejected. Not once or twice, but continuously for seven years. By 1988 he had had enough.

Frustrated, Gangotena called for an emergency meeting of the CPU Foundation. It was clear they were not going to get the government's approval. So, he rallied the troops and decided to go ahead anyway. They started purchasing ads in the media to recruit students, and almost 4,000 prospects showed up. But most of them did not register when they found out the university was not approved by the state. Eventually, 150 students took an admission test and 132 were admitted. It was a small start, but a start after all.

Fifteen days before classes began, however, they lost the facilities promised to them by the Guapulo convent, so they scrambled to rent a 500-m$^2$ house to welcome 132 students and nine teachers. They began transforming the house into an improvised campus: some members donated lumber to make the desks and others donated chairs. The kitchen became the physics and chemistry laboratory, and the basement became a restaurant called The Dragon's Inn.

Today, the University of San Francisco de Quito (USFQ), or San Pancho, as it is affectionately known, is by many accounts the best university in Ecuador.* It no longer operates from a house. It has a beautiful campus that includes a small lake with an Asian-inspired pagoda. The university that grew out of Gangotena's passion, tenacity, and frugality was only officially recognized by the government in 1995, three years after graduating its first cohort of students. It has also proven to be an extremely adaptative organization, changing campuses multiple times as it has grown from its humble origins into its current status. San Pancho is a testament to the idea that universities are more a network of people than a place.

---

* It is number one in the QS ranking (801–850) (www.universityrankings.ch/results?ranking=QS&q=Ecuador) and the nature index (www.nature.com/nature-index/country-outputs) Ecuador and number two in the Times Higher Education (THE) ranking (1201–1500) (www.timeshighereducation.com/student/where-to-study/study-in-ecuador).

But to us, it represents a point of reflection. It is sobering to realize that the best university in Ecuador did not start from a multi-million-dollar government megaproject, but from a handful of passionate academics who were forced to defy the way in which their government did things. What if Yachay's US $1 billion had been used to endow San Pancho and other well-performing local universities? Would that endowment have helped educate more Ecuadorian students? Would it have produced more and better research? We can only speculate. But we should remember that the brightest flames are not always the largest. They are often those that are best at challenging the darkness.

# 12.

# *The Laws of Knowledge*

More than 2,000 years ago Archimedes found a simple way to prove that $\pi$ had to be a number between three and four.

Imagine a circle with a radius of length one. That circle fits perfectly inside a square of side two. It can also perfectly enclose a hexagon with sides of length one. Now, if you compare the perimeter of these three shapes you have to conclude that the perimeter of the circle has to be shorter than that of the square and larger than that of the hexagon. So, we can use these simple drawings to prove that $\pi$ must be smaller than four and larger than three.*

For centuries, scholars used Archimedes' idea to approximate $\pi$. They surrounded the unit circle with octagons, dodecagons, and eventually with polygons made of hundreds of sides. With each iteration, they grew closer to the mysterious number, but at an increasing cost.

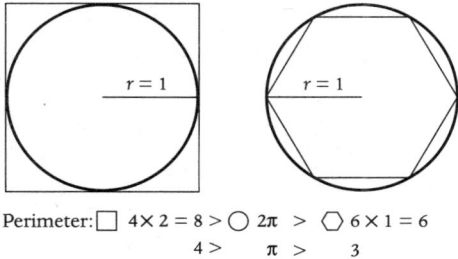

Perimeter: ☐ 4 × 2 = 8 > ○ 2$\pi$ > ○ 6 × 1 = 6
              4 >      $\pi$ >        3

*Figure 25: By bounding the perimeter of a circle with a square and a hexagon, Archimedes showed that $\pi$ must be a number between two and three.*

* There is an excellent video summarizing this story on the YouTube channel Veritasium.[191]

The German-Dutch mathematician Ludolph van Ceulen spent his life enthralled in a quest to approximate $\pi$. Van Ceulen spent years calculating a polygon with over 2 billion sides. This led to a 1596 book where he published $\pi$'s first twenty digits. But he didn't stop there. He eventually expanded his work to the thirty-five decimals that would be engraved on his tombstone.

But this was probably the last time that someone dedicated their life to approximating $\pi$. A few decades later, a twenty-three-year-old Isaac Newton figured out how to estimate many decimals of $\pi$ in a matter of days. Instead of approximating a circle with polygons, Newton combined an infinite sum of increasingly smaller terms with the calculus he had also invented.* Today, there are even better methods to approximate this famous irrational number. Scientists and mathematicians are aware of hundreds of formulas to approximate $\pi$. In the light of this, Van Ceulen's work seems futile. It was the last step of the learning curve that Archimedes had started. Newton's work leaped into a learning curve that escaped Van Ceulen's imagination.

About 50,000 words ago, I promised to take you on a journey through the principles governing the growth, diffusion, and valuation of knowledge. I motivated this journey with stories of failed attempts to build cities of knowledge. But this book is not about those cities. It is about the scientific principles that govern the growth and accumulation of our planet's most valuable non-substance. So, to conclude, I would like to summarize these principles, not by retelling the same stories, but by sharing some new ones.

The story of Van Ceulen and Newton is a clear nod to the first principle: the principle of time. The one involving the work of psychologists like Thurstone, engineers like Wright, and economists like Rapping, Christensen, and Henderson. The line from Archimedes to Van Ceulen is a learning curve with strongly diminishing returns.

Van Ceulen did not lack determination. But to calculate $\pi$ you

---

* Newton's method is based on the series expansion of $(1+x)^{1/2}$.

need more than elbow grease. You need ideas. The first part of this story should remind you of Thurstone's typing class or Wright's aircraft manufacturers. Van Ceulen climbed a known learning curve as high as he could. But the second part of this story, the one where Newton comes in, should remind you of Christensen's disruptive innovation. Newton is the disruptor in this story. Not only did he produce a technique that was superior to the one proposed by Archimedes, he proposed a technique that could be made much better. That gap, the one between the sweat of Van Ceulen and the brilliance of Newton, is the one that explains the exponential growth of knowledge described by Gordon Moore and that is at the heart of Christensen's idea of disruptive innovation. It is that succession of Schumpeterian waves of creative destruction that are a testament to the value of ideas. But even ideas are not enough.

What Rebecca Henderson's work has shown us is that even when companies are aware of a new idea, they can struggle to adapt. Jumping to a better learning curve is hard when this change requires architectural knowledge – knowledge that requires rewiring an organization. Blockbuster's executives clearly knew about online DVD rentals when they refused to buy Netflix for $50 million. But this was an idea that was hard for them to implement in an organization that was not built around mailing DVDs directly to consumers. The story of Amazon and Barnes & Noble is another case in point. It is not that the executives of Blockbuster or Barnes & Noble lacked determination, but that adopting an idea as an organization involves challenges that transcend the learning abilities of individuals.

The 'first law' of knowledge unfolds across time. It is not a neat one-liner, like Newton's first law of motion, but more of a mnemonic device that we can use to remember a number of interconnected principles involving learning, experience, and forgetting. Still, now that we know its contents, we can try to formalize it into a single paragraph:

Knowledge growth is governed by principles rooted in learning curves, which describe how individuals and organizations improve by accumulating experience through repeated practice. Learning curves show diminishing returns but can combine into broader experience curves – spanning multiple iterations and innovations – resulting in the exponential growth observed in Moore's law. The process of innovation, however, often involves disruptive dynamics, where newer, initially inferior technologies overtake incumbents through architectural shifts that require rewiring organizational networks of people, tools, and tasks. Yet, this growth is not guaranteed, as collective forgetting – driven by a loss of communicative and codified knowledge – can erode hard-won expertise, particularly when production ceases or slows down.

Unfortunately, the first law lives in apparent isolation, since it doesn't tell us how the growth of knowledge is influenced by geography or the presence of related activities. That requires the second principle. Still, the first law has some important implications.

One of them is the difficulty of competing against incumbents when you lack a disruptive innovation strategy. If you lag behind others in the same learning curve, you are unlikely to win. The best you can do is play catch-up. And even if you manage to catch up, you will end up competing for slim margins. We probably would not be talking about Newton if he had dedicated his life to one-up Van Ceulen by calculating a thirty-sixth digit of $\pi$.

But the first law also tells us that it is important to jump early into a new technology. Back in the 1890s, putting any working engine on a horse carriage made you a car manufacturer. That didn't guarantee your survival, but it meant you at least had a 'horse in the race.' Thirty years later, manufacturing cars meant competing against the powerful assembly lines of Henry Ford.[192] The first law tells you that technological change opens and closes windows of opportunity each time two learning curves are getting ready to cross. In a world where many people are looking for new opportunities, being too hesitant might mean being too late.*

---

* This idea is related to the work of Brian Arthur, who has long studied increasing returns and their ability to generate monopoly power (first-mover advantage).[193,194]

The first law also has some important systemic implications. These are related to Moore's ideas, since they involve situations in which we should not care about who wins but about the overall ability of the system to learn. Sustainable energy production and consumption are a key example. We have learned from Nordhaus and Fouquet that the cost of illumination has fallen enormously over centuries. If we put our planet first, we should not care about which company succeeds at developing the best and most affordable renewable energy technologies. We should care that at least one of them does.

Many renewable energy technologies, such as electric batteries and solar panels, follow experience curves similar to those of artificial illumination. During the last decade the cost of electric batteries dropped from about $1,400 per kilowatt-hour to a few hundred.[196] Similar experience curves are found in wind and solar.[197] This is not because Bell Labs has continued to dominate the production of solar cells since they introduced them in the 1950s, but because they've long since passed the baton to other companies. This part of the first law tells us that we can expect many of the things we care about – such as renewable energy production – to improve when markets are open and welcoming to new entrants. The policy implication is that markets need to be open enough for David to be able to beat Goliath. As long as that is the case, the system can continue to learn.*

Still, there is also a flip side to the first law: the part focused on forgetting. Knowledge satisfies an 'if you don't use it, you lose it' principle. Remember the quixotic journey of Florian Kaps to save Polaroid? Or the failure of Lockheed's L-1011 aircraft? In our century, there is an ongoing debate on whether 'the West' has lost its ability to build key pieces of infrastructure like nuclear power plants. This

---

Keun Lee and Franco Malerba have also long discussed windows of opportunity in the context of successive learning curves.[195]

* This is related to Karl Popper's idea of the open society, which is basically the idea that a good system of governance is the one that can get rid of bad leaders.

is an idea that would be compatible with the notion that knowledge is lost when you fall out of practice.

In the two decades from 2001 to 2020, the G7 economies* introduced one new gigawatt of nuclear power.[198] This is a sharp drop from the 1980s, when they introduced eighteen new gigawatts in a single decade. Certainly, nuclear power plant manufacturing is a story with many angles, but I think we can all agree that anyone wanting to build a nuclear plant would prefer to hire a crew that has at least built a few.

The 'if you don't use it, you lose it' principle also applies to other technologies that are key for sustainability, such as the construction of new subway lines. Cities that have built subway lines continuously during the last twenty years, such as Shenzhen, Santiago de Chile, or Chongqing, have enormously expanded their networks. Cities that were once pioneers, like Boston, have struggled to complete a few antiquated lines.† They are now back at the beginning of the learning curve, this time with a more complex regulatory environment and a more intricate underground. The moral of the story is that there are critical capacities, like building subways and nuclear plants, that a country should keep active for strategic reasons. The 'if you don't use it, you lose it' principle is a key corollary of the first law, and it means that economies need to be careful about interrupting work on strategic activities.

The second law was our way to organize the principles governing the diffusion of knowledge. This law was separated into two parts: the principles governing the diffusion of knowledge across space, and the principle of relatedness, which governs the diffusion of knowledge among activities.

---

*An intergovernmental political and economic forum consisting of Canada, France, Germany, Italy, Japan, the United Kingdom, the United States, and the European Union as a 'non-enumerated member.'

† Boston's Green Line Somerville extension was a notorious urban infrastructure project that, by some accounts, started in the 1980s or 1990s. It only completed the extension of a surface trolley in the early 2020s.

The diffusion of knowledge across space is a principle that is by all means cinematographic. It includes rags-to-riches stories of ambitious migrant entrepreneurs such as Samuel Slater. So, if you indulge me, I would like to share with you another movie-like story illustrating this principle.

In 2025, we know the city of Donetsk as one of the epicenters of the war between Russia and Ukraine. You may also know that, for more than a century, Donetsk was an important center for the production of iron and steel. But what you may not know is how Donetsk came about, and how its history is linked to that of a Welsh entrepreneur by the name of John Hughes.*

★

John Hughes was born in 1814 in Merthyr Tydfil, at the time one of the richest iron towns in Wales.[199] His father was the head engineer of the Cyfarthfa Ironworks, where John started his brilliant career. His patents in armor plating allowed him to purchase a shipyard by the age of twenty-eight and a foundry when he was thirty-six. In the mid-1850s he moved to London and won worldwide acclaim for the iron cladding of warships. This caught the attention of Imperial Russia. Hughes was invited to plate the Constantine naval fortress the Russians were building in Kronstadt, on the Baltic Sea, and, later, to take over a concession to build an ironworks operation in the southern reaches of the Russian Empire.†

With this assignment in mind, Hughes returned to England to form a company focused on settling the land that a century later would be known as the city of Donetsk. But Hughes did not travel alone. He packed eight ships with 100 ironworkers and miners, mostly from South Wales. The materials were imported from England too, via the Azov port of Taganrog. The conditions were brutal.

---

* Not John Hughes the American film director and producer known for movies such as *The Breakfast Club* and *Home Alone*.
† A concession granted to Prince Kochubey in 1868.

They had to haul the materials by ox for 100 km through muddy terrain. Still, on April 24, 1871, Hughes' new enterprise smelted its first piece of iron, and by 1874, the blast and puddling furnaces were working continuously, producing 150 tons of pig iron* per week and about 8,100 annual tons of rail.

Hughes' ironworks gave rise to the fast-growing city of Hugheskova, or Yuzovka. The metalworks provided a hospital, schools, bathhouses, and even an Anglican church. By 1913, Yuzovka produced 74 percent of Russian iron.[†] But in 1917, the foundry was seized during the Bolshevik revolution. Yuzovka was renamed Stalino in 1924 and Donetsk in 1961. Yet, despite the incessant renaming, 'Yuzovka' continued to honor Hughes' legacy, growing as a hub for metallurgy with a metropolitan-area population of over 1 million.

Hughes apos story differs from that of Slater in some key aspects. Slater left England secretly at a young age. Hughes spent most of his life in the UK and left only after becoming a successful entrepreneur.

Also, Hughes' story involves an industry that was more mature. When Slater arrived in the United States, he was the only person in America with any experience running a water-powered cotton-spinning mill. He only needed a few men to run the first mill. Hughes, however, was transferring an operation that required the specialized knowledge of dozens of men and had to pack eight ships to carry that operation. He also competed for the concession with other potential providers, meaning that the knowledge he was bringing was more widespread.

Moreover, Hughes moved to a place where he did not speak the language. That probably motivated him to bring a large team, since he could not rely fully on locals for management. This emphasizes the notion that knowledge is contained in organizations. Hughes' ships were packed with architectural knowledge.

The story of Yuzovka also differs in important ways from that

---

* Pig iron is a crude form of iron used as an intermediate input for the production of steel. It is made of ingots molded in the sand through a branching structure, that makes the ingots look like nursing piglets.
† About twenty years after Hughes' death.

of Yachay. Unlike Yachay, which was an attempt to build a city that wanted to excel at everything, from biotechnology to robotics, Yuzovka was focused solely on metalworks. Its choice of location was also strategic. It was near coal and iron deposits,* two key resources for the production of steel. These factors made Yuzovka a more viable effort than Yachay. To succeed, Yuzovka had to reproduce the narrow success that Hughes had already achieved in the United Kingdom. Yachay had to build a diversity of activities that went beyond what anyone in Ecuador had ever achieved.

This distinction brings us to the second part of the second law: the principle of relatedness, the law focused on the movement of knowledge among activities. In some cases, like Yuzovka's focus on iron and steel, relatedness seems obvious. But it is important to keep in mind that this is not always the case. To illustrate this idea, let me tell you a story from when I brought a team of MIT, Harvard, and Tufts students to Grand Rapids, Michigan to visit Steelcase in 2016.

In the fall of 2016, I was running a course on applied data visualization. I had teamed up with Steelcase, a large office-furniture manufacturer that had been in business for more than 100 years.† During the semester, students had to design a data visualization solution tailored to the needs of the company. This required us to learn about the information Steelcase collected, shared, and cared about in the course of its operations.

---

* The idea that there was iron in the region was rather new at the time. As J. N. Westwood explains in his article on John Hughes and Russian metallurgy:[199] 'The celebrated French mining specialist Le Play had surveyed this part of Russia in 1837 and had declared that although the coal was excellent there were no iron-ore deposits. The word of this oracle was unchallenged for decades. In fact, a government metalworks, which was established at Lugansk, imported iron for processing from the Urals – a procedure which made its output prohibitively expensive. It was not until the 60s that the government's Mining Department realized that local ore could be exploited in southern Russia.'

† Steelcase's name was inspired by what was once its signature product: a fire-proof metal wastebasket. Keep in mind that people used to smoke inside offices and would discard cigarette butts in paper bins.

That visit taught us plenty about Steelcase's corporate culture. This is a company that values storytelling as a mechanism to communicate and preserve what they refer to as tribal knowledge. Our contacts at Steelcase showed us several factories and introduced us to many foremen, factory floor workers, and executives. At the end of the day, we had a better picture of what metrics they cared about and why. These metrics spoke volumes about what matters in office-furniture manufacturing.

The story of Donetsk overemphasizes the idea of value-chain relationships, since pig iron is an input for the production of steel. But because value-chain relationships are so intuitive, they can warp our sense of product or industry relatedness. In fact, input–output thinking is not always correct. In the case of Steelcase, the two metrics that they followed more closely were not the price of inputs, such as sawn wood or sheet metal, but 'shorts' and 'splits.'

To understand what these two numbers you need to know that Steelcase is a real-time manufacturing company. That means they manufacture something only *after* making a sale. So, everything they do on the factory floor contributes to making an item they've already sold. The advantage of this is they do not need to keep unsold products sitting in a warehouse. But as you can imagine, this means running a spectacularly tight operation.

At Steelcase, each sale starts a countdown. In a few days, all of the items in an order need to be loaded onto a truck. If an item is missing, and the truck leaves, that shipment becomes a split, meaning that they need to deliver some of the items in a second shipment. If they are running behind on a few items before the truck leaves, they count that as running short. Shorts anticipate splits, but splits are the headaches they are trying to avoid. They are the fine line dividing profits from losses.

But for us, what splits and shorts tell us is that large-scale furniture manufacturing is not closely related to the business of sawing wood. Steelcase's metrics reveal that it is a logistics-intensive operation. Its Grand Rapids location makes sense, not because of Michigan's forests but because of Michigan's tradition of logistics and manufacturing.

In hindsight, this might look obvious. If you've ever been to Ikea, you know that what you see in the showroom comes in an airtight box. The problem of selling large volumes of furniture is not that of putting together steel rods and wood, but that of moving furniture across the country. Steelcase, Ikea, and other furniture manufactures design items that are easy to pack, assemble, and transport. This makes furniture manufacturing related to other logistic-intensive business, such as vehicle manufacturing, which has a long tradition in Michigan. Furniture is technically downstream from the production of sawn wood, but in terms of relatedness, it is in the purview of industrialized economies, like Italy, Germany, Poland, China, or Turkey. What relates activities to places, or to each other, is sometimes less obvious than what value chains would have us think.

The Steelcase example should invite us to think creatively about relatedness. This is because relationships based on shared knowledge can be quite different from value-chain relationships. This should make us cautious about naïve downstream diversification policies, which are abundant in the developing world.

In Chile, there are plenty of people who believe that Chile's vast lithium-ore deposits provide the country with an advantage for electric car battery manufacturing. But the beautiful salt lakes of the Chilean desert, from where lithium ore is extracted, look quite different from the dense urban neighborhoods in Shenzhen where hundreds of thousands of researchers are working on the next generation of battery technologies. There is in fact evidence that, even in the context of value chains, relatedness might operate more effectively against the grain,[200,201] meaning that the shortest path from a mine to a finished product is not the one flowing downstream from the mineral to the metal, but the one moving upstream from the ore to the equipment needed to extract it.

There are some good reasons why upstream movements might make more sense. Any company that is involved in a value chain depends heavily on its customers. Moving downstream means competing directly with the customers you depend on financially; customers who might be unwilling to continue buying inputs from a company that turned into a competitor. Moving upstream, on the other hand, means cutting off your suppliers. Such upstream movements are a common business strategy. Broadcasters in the United States have used it repeatedly, from Turner Broadcasting creating the Cartoon Network* to Netflix more recently developing its own movies and series. In France, the video game company Ubisoft began selling third-party games before they began making their own. But what I think might be the most riveting story illustrating upstream movement is that of Lenovo becaming a personal computer manufacturer. A story that brings us back to Zhongguancun.

*

In the 1980s, personal computers were a fast-growing product. But in China their adoption was limited by the fact that computers did not run in Chinese. In *Zhong Guan Village*, Ning Ken tells the story of how Lenovo† grew by exploiting this window of opportunity.[142]

In the mid-'80s, Lenovo's founding figure and leader Liu Chuanzhi convinced Ni Guangnan, an accomplished scholar and creator of the LX80 Chinese character card, to join him in a commercial venture. Back in the mid-'80s, personal computers were not powerful enough to afford doing character conversions directly through software, so people would install physical cards for that task. Guangnan's LX80 was one of several Chinese character cards designed for that purpose, but it had a neat auto-complete feature that doubled a person's typing speed. That made it an interesting alternative for Lenovo, which grew by selling several generations of these cards.‡§

* This was of course not a straightforward business operation: it included among other things buying assets from Hanna-Barbera while starting to make cartoons on their own.
† Originally called Legend.
‡ The Mk I in 1985, followed quickly by the Mk II and the Mk III. In 1987 Lenovo sold 6,500 units, solidifying itself as an emerging tech leader.
§ Lenovo's success, however, caught the attention of officials who were suspicious of private sector activities. They accused Lenovo of profiteering and fined

Chuanzhi understood that computers were not going to need a dedicated hardware component to deal with Chinese characters forever, so he devised a strategy to transform the company into a more central player in the business of personal computers. Lenovo had developed a good commercial arm during the 1980s, but it still depended strongly on its card working on computers manufactured by other suppliers. That motivated Chuanzhi to visit Shenzhen in search of a personal computer that Lenovo could sell together with its card. After trying a few brands, Lenovo became a distributor of AST, a personal computer manufacturer out of Irvine, California.

Lenovo successfully expanded AST's sales in China, learning in the process about the technical and commercial aspect of the business of personal computers. But behind the scenes, it began working on their own personal computer: Lenovo's model 286. In March 1990, the Lenovo 286 received a permit for production, which enabled Chuanzhi to execute the next part of his strategy. Overnight, he instructed his sales force to stop selling the AST 286 and sell instead the Lenovo, a lower-priced alternative that allowed Chuanzhi to swallow up his supplier and complete his vertical integration strategy. There was not much that AST could do to retain the Chinese market. By eating its own supplier, Lenovo went from being a manufacturer of a computer peripheral to a key player in the world of personal computers.

In principle, the second law seems simple: it is about knowledge moving across geographies, social networks, and cognitively related activities. But underneath that simplicity there is nuance. On the one hand, there are processes governing the formation

---

it 1 million yuan, which was roughly Lenovo's entire net profit. Chuanzhi fought back, albeit with a light touch. After trying to meet the director of the Bureau of Commodity Prices several times at his office, he surprised him at his home, on a Sunday, while he was having dinner with his family. Chuanzhi came with a female colleague who delivered the company's message with kindness. The visit made all the difference, leading to a reduction of the fine from 1 million to 600,000 yuan.

of social and professional networks, which are themselves areas of scholarly study. These network formation processes shape economic transactions, as the sociologist Mark Granovetter famously argued when he advanced his theory of embeddedness.[86,87] In a nutshell, this is the notion that many economic transactions flow through networks that were formed for reasons unrelated to those economic activities, such as high-school friendships and kin relationships. On the other hand, there is also nuance to the fact that some knowledge flows are primarily intergenerational, as we learned from the migration of German chemists in the 1930s or from the patterns used by classical music composers.[88,93] This fact puts value on apprenticeship models, but also tells us about the limited speed of knowledge diffusion.

There is also nuance in the fact that links among economic activities can be somehow counterintuitive, as we saw in the Steelcase example. What may seem obvious in principle, like moving into a downstream product, might be a naïve strategy in a world where the similarity between activities is dominated by knowledge relationships.

But what might be strategically more important is to understand that while relatedness is a natural economic force, we should not let it define our development strategy. There are three key reasons for this.

First, there is mathematical evidence that focusing solely on the most related activities is a sub-optimal diversification strategy.[202] A focus on related activities can trap an economy in a local optima, which can be hard to escape without attempting some 'longer jumps.' A second reason is that relatedness is only one of the many factors constraining an economic diversification strategy. As a matter of fact, recent advances in the academic literature focus on optimizing portfolios of efforts to enter related and unrelated activities. These efforts use relatedness as one of many factors.* And a third reason is that these laws of knowledge provide a philosophically positive

---

* Viktor Stojkoski and I have been working on this approach for more than a year now.

description of the world, meaning they tell us how the world works and not how we should behave. To understand this, consider the following analogy.[203]

Imagine trying to navigate a sailboat. You can think of the principles in the first and second laws as the direction and strength of the wind. They tell us how fast knowledge grows and in which direction. But sometimes, a sailboat needs to navigate into the wind. For many developing countries, thinking strategically means thinking about wind-defying strategies – strategies to develop product, industries, and occupations – that would be hard to grow organically.

But while the second law in general, and the principle of relatedness in particular, brings us closer to the idea of non-fungible knowledge, we still need ways to estimate the value of knowledge agglomerations. This is where the third law comes in. A law that tries to put all of the pieces of the infinite alphabet together using a single recursive formula. But what is this formula really capturing? Maybe, the easiest way to communicate this, is to share with you a story separating talent from opportunity. A story trying to illustrate that opportunity is not determined by a single factor, but could be approximated by measures that attempt to capture 'all of the above.' This is the story of Rene Rios and Walt Disney, two extremely talented individuals who lived very different lives.

★

Walt Disney's story is well known, so I will review it quickly. He was born in Chicago in 1901 and raised in Missouri. After taking night classes at the Chicago Art Institute, he moved back to Kansas City, where he began working as a cartoonist for a local newspaper. In his late teens he started a company with his friend and lifelong collaborator Ubbe Iwerks.* But their business did not take off and Disney had to take a job at the Kansas City Film Ad Company. After a short

* Iwerks – Disney Commercial Artists.

period in advertising, Disney became an entrepreneur again by creating the Laugh-O-Gram animation studio. Laugh-O-Gram created some successful shorts, which allowed Disney to hire new animators, including Ubbe Iwerks. Together, Disney and Iwerks moved to California where they created their first hit, Oswald the Lucky Rabbit under a contract with Universal Studios.

Things were going well until Universal told Disney that they would poach his animators if they did not reduce their production costs. But the threat backfired when Disney and Iwerks responded by creating Mickey Mouse.

Mickey Mouse shorts launched Disney's career into orbit and got him a 1932 Academy Award. But he had bigger plans. In 1934, he began working on the production of a fully animated feature film. A ludicrous idea at a time when production costs restricted animation to shorts.

Making *Snow White and the Seven Dwarfs* meant Disney was betting his entire studio on a single multi-year project. To complete it, he secured a loan by showing incomplete versions of the film to bankers. But the bet paid off. Once the movie was out it became an instant success. The animated movie was received with a standing ovation when it premiered at the Carthay Circle Theater on December 21, 1937. In today's money, *Snow White* made about a billion dollars of box-office revenue and resulted in an Oscar that was given together with seven little golden statues.

That windfall made Disney unstoppable. His empire grew from movies and TV shows, to music, magazines, theme parks, and merchandise. By the end of his life, he was moving into real estate by designing futuristic cities. The Experimental Prototype Community of Tomorrow, or EPCOT, was a living laboratory designed to test urban planning concepts and technologies. Disney's death, however, meant EPCOT was reimagined as a theme park. Still, we can think of it as a monument to Disney's outsized creativity and ambition.

René Rios, or Pepo, was also a creative genius. He was born in

the Chilean city of Concepción, only ten years after Walt Disney.*
Pepo created Condorito, arguably the most famous Chilean cartoon
character. Like Disney, Pepo showed talent from a young age. He
published his first newspaper cartoon at age seven and created an
exhibit in a candy store at age ten. After dropping out of medical
school,† Pepo moved to Santiago to study fine arts and began drawing political cartoons for newspapers and magazines.

It was at that time that Pepo had a brief encounter with Walt
Disney. During the Second World War, Disney visited Latin American on a tour organized by the US Department of State. After the
visit, Disney released the movie *Saludos Amigos*, an animated film
including the story of a Chilean airplane called Pedro struggling to
cross the Andes.

It is rumored that Pepo created Condorito in reaction to this
movie. He believed that Pedro the airplane did not represent the
Chilean character well. Unlike Pedro, who was a well-mannered and
hard-working airplane, Condorito was a mischievous anthropomor-
phized condor who become popular throughout Latin America.

Like Mickey Mouse, Condorito was a versatile character. He
could be a fireman in one comic strip and an astronaut in the next.
He was also accompanied by a lively cast of characters who populated the fictional town of Pelotillehue. They include Condorito's
love interest Yayita, his nephew Coné, and friends like Don Chuma,
Huevoduro (hard-boiled egg), and Comegato (cat eater). Condorito
also had enemies, such as Pepe Cortisona, who would compete for
Yayita's love, and Doña Tremebunda, Condorito's mother-in-law.
But unlike Mickey Mouse, Condorito was never able to successfully
transition from print to screen. Efforts to animate Condorito during
the twentieth century were often short lived, poorly executed, or

---

* This section builds on a few online sources and on Luis Yáñez's 2020 book *Pepo es de Conce*.[204]
† In Chile, medicine is a seven-year-long undergraduate degree, so this would be the US equivalent of dropping out from a pre-med bachelor's as a sophomore.

cancelled. Eventually, a team did complete and release a 3D animated Condorito movie in 2017, eighty years after *Snow White*. A movie with a budget of $8 million that collected about the same amount.

Both Condorito and Mickey Mouse grew from the creative genius of talented graphic artists. Pepo, in the case of Condorito, and Disney and Iwerks in the case of Mickey Mouse. But despite the fact that the Pepo and Disney stories start similarly, they diverge enormously by the end. So, one is left wondering what would have happened if Disney and Pepo had traded places. If Disney had been born in Concepción and Pepo in Chicago.

Both Pepo and Disney moved to a larger city looking to develop their skills. They both studied art and began their careers drawing comic strips for print publications. And they both became successful in their own right. But that's as far as the similarities go. As they both grew out of their shells, Disney became a magnet for diverse forms of talent, whereas Pepo had a much more modest roster of collaborators.*

Once he established himself in Hollywood, Disney was able to collaborate with talented animators, writers, musicians, actors, filmmakers, and innovators. By the 1950s, Disney was building theme parks and had established a long collaborative relationship with Salvador Dalí. Santiago de Chile in the 1940s was certainly no Hollywood. But that is exactly the point. The point is that what differentiates economies is not a thing or two, but so many things that we need measures that are good at capturing 'all of the above.' Measures that are, in principle, agnostic about what counts, but that in practice can count what matters.

Economic complexity measures try to do this by using

---

* This story is clearly not about basic living conditions. Pepo was born in an upper-income family. His father was a doctor who participated in the founding of a university. Disney grew up in a low-income family with four other siblings. So, when it comes to basic living conditions, Pepo probably had a more comfortable early life.

fine-grained data on the economic activities that are present in each country or city. This is not because the activities are themselves the factors, but because if activities require many factors, their geographic distribution will act as a 'barcode' or 'fingerprint' telling us about their factors. These factors could be quite diverse. They could represent the availability of talent, like the one expressed in Hollywood's burgeoning movie industry in the 1920s. But they could also represent more subtle cultural factors. When Disney met Pepo and other Chilean graphic artists, he critiqued their drawings.[204] The Chilean artists did not like this (they may have been looking for praise instead of feedback). This may be one of the reasons why Pepo reacted by creating a character that was the opposite of the one created by Disney. But had the Chileans understood Disney's culture of feedback differently, they might have been able to develop a collaborative relationship with the American tycoon. Also, maybe their Condorito-like behavior might be one of the reasons why some of the many efforts to move into animation failed, since animation requires building larger organizations than printing a children's magazine.

To our point, what matters is that a measure of complexity based on the presence or absence of economic activities across a wide range of geographies can be accurate while still being agnostic about the nature of the factors. As long as the factors affect the presence or absence of the activities, they will be captured indirectly. No matter if these factors are cultural, institutional, or even geological (e.g., landlocked countries having a limited production of saltwater fish).

One final analogy that we can use to try to understand this idea is to think of Disney and Pepo as grape seeds that need fertile soil to grow. Soil that contains a complex combination of nutrients that will determine the quality of the wine. As any winegrower knows, you cannot grow good wine just anywhere. To grow the vine of knowledge you also need good land.

★

One of the biggest disappointments of growing up is learning that the universe is finite. There is something romantic about an infinite and eternal universe. How can 'everything' be so limited in terms of volume and time? But even Einstein had to concede. After discovering his famous equation (not $E=mc^2$, but the one for general relativity)[*] he introduced a term to his model so that the universe would not expand or contract. Einstein's cosmological constant made the universe static, and therefore eternal. But he would eventually call this his biggest blunder. Today, we have many numbers to describe the finite size and age of the universe. The age of the universe is estimated to be around 13.8 billion years. Its diameter, about $10^{26}$ m. That is by all means large. Our solar system's diameter is a few billion km, meaning that the universe could fit a trillion solar systems side by side. Our universe is also home to a finite number of particles, of the order of $10^{80}$. That's a one with eighty zeros. A big number, but one that we can still write on a single sheet of paper.

But maybe the way to regain some of that lost romanticism is to think of our universe not in terms of age, size, or particles, but in terms of possibilities.

Consider an alphabet with 100 unique letters. Now, imagine a blank sheet of paper, large enough to fit 1,000 characters.[†] Now, ask yourself, how many possible sequences of characters you can type in that page? Since for each character you have 100 options, and you can do this independently 1,000 times, that means $100^{1,000}$ or $10^{2,000}$ possibilities. That is still a finite number, but one that makes the number of particles in the universe look tiny. Our universe might be finite, but when it comes to possibilities, even something as simple as a blank piece of paper can present us with 'endless' opportunities.

This is certainly not a new idea. It is the point made by Jorge Luis Borges in 'The Library of Babel', which in turn borrowed this idea

---

[*] This is an equation connecting the geometry of space–time with the density of energy and momentum: $R_{\mu\nu} - \frac{1}{2}Rg_{\mu\nu} = 8\pi GT_{\mu\nu}$.

[†] That's about half a sheet of A4, or letter paper.

from Kurd Lasswitz's 'The Universal Library'. But I am not making this point because it is new, but because it is powerful. It is a way to recover the sense of wonder that we lose when we learn about our finite universe.

In a deep sense, knowledge lives in this second universe. This larger universe made of options and possibilities. A universe where we can write an infinite number of songs with a finite number of notes. A universe that hides an infinite number of stories in a few pages.* Thurstone's typing class students may make us think of knowledge as something finite, but infinity is not about trivial things, such as a student's typing speed, but about the complexity of new combinations. New combinations that are made possible by people, because we are the ones who carry knowledge. Each of us carries only a few letters, but together, we get to move those letters around and discover new letters and combinations. It is through the virtue of our lives that our feeble minds and weak bodies give rise to something larger than our minds. Something beyond our comprehension. The most valuable treasure of our species: knowledge's ever-growing infinite alphabet.

---

* I am certainly using infinity here metaphorically, not literally, to indicate a very large number of combinations, much larger than the number of particles in the universe.

# Afterword

When I met Charlie at the lounge of Chicago's O'Hare airport I was thinking about writing a book.

That book idea went through many names. At first, I had the horrible idea of calling it *Monkeys Playing Lego*. Eventually, I sold it under the title *Crystallized Imagination*. But halfway through the writing process my editor told me that the title was too dreamy. I had to concede. *Crystallized Imagination* – the idea that the objects we make and transact are physical vehicles for the practical uses of our knowledge – was a strong concept wrapped in a fluffy name. The idea still survived in the book, but as a chapter instead of its spinal cord. So, I had to find another way to put everything together.

For a brief period of time *Crystallized Imagination* become *Out of Our Heads*, but that title was killed because it alluded to madness. Then, the book became *The Bit and the Atom*, and eventually, during a work trip back to Chile, in August of 2014, I thought of the final title: *Why Information Grows*. After searching for this phrase on Google with quotation marks, and noticing it had almost never been used, I was sold.*

Like this book, *Why Information Grows* (*WIG*) also explored the difficulties of accumulating knowledge. But it approaches that problem from a different perspective. It starts from the idea that 'crystals of imagination' need knowledge. Technically, more knowledge than what a single person can know (the idea of a 'personbyte,'[205] which measures knowledge at the collective scale in units of what individual can accumulate). The main question in *WIG* then was why some societies are better than others at accumulating many personbytes of knowledge. That implied the need for some social

---

* I included a screenshot of that at the end of *WIG*.

glue. Something that can keep us cooperating beyond a naïve mathematical model of incentives. Something that we can not only understand but feel. That led me to the literature on trust.

Trust, I argued, changed the size of the social and professional networks that people form. From lots of tiny clusters in low-trust societies, to giant hairballs of cooperation. Societies lacking trust struggle to accumulate many personbytes of knowledge because they cannot easily build the networks they need to accumulate it.* Accumulating multiple personbytes of knowledge requires large networks of cooperation that are easier to build when the cost of interactions is lower, which is in societies where trust is higher.

The point was that rich and prosperous countries had this social glue. Trust was a 'parameter' that could control the size of a knowledge 'container,' and therefore, explains the difference between a society that is able to grow vast professional networks, with specialized people and a society fragmented into tiny networks, where people compete on the same activities.

I still stand by these ideas. They are simple, powerful, and yet can be infinitely unpacked. But the trust and personbyte story was the second part of *Why Information Grows*. The first part was about the physics of information. Probably that's where I lost most of my readers. The chapters made the point that, at a very fundamental level, the universe is made of things, movement, and configurations, and that this last part may be the most important. This point has been made by many physicists. John Wheeler famously made this point in his essay known popularly as 'It from Bit'.[206] Erwin Schrödinger was at some point so enthralled with this idea that he wrote that life feeds primarily on 'negative entropy,' a remark or which he later had to backpedal.[207]† For much of history, physics was about understanding things in motion. It was the science of mass, energy, and momentum. But during the nineteenth century, motivated by the

---

* Or the 'firmbyte': the knowledge-carrying capacity of an organization.
† See note to chapter 6 in Erwin Schrödinger's *What Is Life?*

study of heat, scholars like Ludwig Boltzmann began to figure out that configurations of matter had real macroscopic consequences.*

For the better part of the last century, we used measures of configurations, such as Shannon's information, to help us communicate through digital networks. But entropy is not just a convenient quantity for building telecommunication networks. It is a fundamental property of all macroscopic objects. We understand intuitively that things have a size, a weight, and a temperature. The thing is that all objects also have an entropy too. The idea that the fundamental building blocks of the universe are not material, but informational – in a very raw physical sense – is still something that I believe is a key point of departure for modern science. It is an idea that pushes us to accept that there are fundamental questions about the nature and growth that are shared by all scientific disciplines, from fundamental physics to economics and biology.

But after the publication of the book, I had some regrets. In the ten years following the publication of *Why Information Grows* I learned things about the principles governing the growth and diffusion of knowledge that I was not aware when I wrote that book. I also understood the importance of communicating concepts using stories instead of abstract ideas. Moreover, I learned about the importance of having a clear plan before you start writing, since writing *WIG* was an extremely painful process. So, I thought deeply about the book's title and its structure.

I was able to figure out the structure first. It came together for a course I taught at Harvard's master in design engineering program in 2019, and that I have been teaching since. But the title of the course, 'Principles of Collective Learning', was a bit too formal for a trade book. Eventually, one day in my apartment in Toulouse,

---

* That fundamental measure of configuration, or more precisely of how many configurations are macroscopically equivalent – what we physicists call entropy and computer scientists call information – is a formula that estimates the number of yes or no questions that one needs to optimally guess the microscopic state of a system.

Afterword

I had an *aha!* moment while talking with Anna. This time, the title survived till the end.

So, if you are wondering if *The Infinite Alphabet (TIA)* is an attempt to rewrite *Why Information Grows*, then the answer is both yes and no. It is clearly not an effort to correct it, but it is an effort to expand it using what I know today. This makes *The Infinite Alphabet* very different to *Why Information Grows*. It has virtually no overlap in terms of style or content. *Why Information Grows* built mostly on concepts, analogies, and abstract ideas. From crashing an imaginary Bugatti to waxing poetic about the difference between apples and Apple. *The Infinite Alphabet* uses fewer thought experiments and more stories. Stories that took years to collect, and that I hope made the ideas presented in *TIA* more memorable and accessible.

These stories were chosen to help the reader learn three laws. The first one is a set of concepts governing the growth and decay of knowledge. How learning curves peter out. How they add up to experience curves. And how their overlapping humps explain changes in the leadership of markets. These ideas have been dressed up in stories: Thurstone's famous typing class, Wright's adventures during the golden age of aviation, the quarrels between Shockley and Moore, and the decaying cost of artificial light and DNA.

But knowledge also moves. In *Why Information Grows* I explained this using a jigsaw puzzle analogy and talked about the difficulties of moving multiple puzzle pieces all at once.[*] In *The Infinite Alphabet*, I have used the heroic story of Samuel Slater to set up the chapter on knowledge diffusion. We have visited Tokyo to learn about absorptive capacity, as Masaru Ibuka and his team developed a magnetic-tape business using a frying pan and a brush. Later, we revisited Slater's story with the heroic experience of the Welsh entrepreneur John Hughes.

But the second law also has a few branches. Knowledge not only

[*] I also mentioned Operation Paperclip, the American plan that brought Wernher von Braun and thousands of other scientists to the United States in the aftermath of the Second World War.

travels along the links of social networks, it also has a geography of its own, a geography we could see by looking at the Italian and Japanese aircraft industry in the years following the Second World War. In Italy, we have learned about Corradino D'Ascanio, the aircraft engineer who designed the iconic Vespa. We have then learned about repeated instances of this same story in Japan. This has brought us a bit closer to the idea of the infinite alphabet, since it has forced us to accept knowledge as a highly differentiated quantity. We were finally back to the idea of configurations.

To introduce the third law, I have explained that it is possible to measure configurations. First, I have tried to explain this using historical 'alphabets.' Alphabets like the ones used in music that go by *Do*, *Re*, and *Mi*. These are alphabets that capture the behavior of physical systems, like guitar strings, and that can help us understand concepts like melody and harmony. I have argued that the idea of breaking up shapes into alphabets is so powerful that it is, in fact, a key building block of quantum mechanics and the language models we equate to artificial intelligence. But since in economic systems we don't yet have a complete description of the forces that 'shape the motion of the economic string,' we must adopt a different approach. A method that we can use to factor out the main modes of the resulting configuration:* measures of complexity, a word that is historically linked to the idea of temperature, and that simply means that a system is be composed of multiple parts.

Together, these three laws or principles are the edifice I promised you in the introduction. An edifice of three floors but infinite rooms. After all, we need an infinite number of rooms to host all of the stories that could be written with an infinite alphabet.†

---

* By being careful about the fact that data on the geography of economic activities mixes units of observation of very different sizes.
† And yes, this is a nod to Hilbert's paradox ☺.

# Acknowledgments

There are many people who supported me through the never-easy journey of writing a book. But there is one who takes precedence over all of them. That is my wife, lifelong companion, and the mother of my only daughter: Anna Sokolovska. Anna knows me better than anyone in this world. She understands the tormented mind of this Chilean-born writer. She has been with me through some of my life's most difficult and defining moments, never leaving my side. I may talk a big game when it comes to discussing the scientific laws of knowledge, but this is not really who I am. The real me is the person who is at home with her, not wanting to leave for work in the morning and dreading every professional trip. This book, and everything I have ever owned or will own, goes to Anna.

But there are also other people who have supported me through this journey. First, I want to thank my editor, Keith Mansfield. Working with him on this book was a delight. His comments were always on point and were essential to achieve the clarity that I hope the book communicates. I am also very grateful for the comments I received from Paul Seabright, who devoured a late draft of this book, provided me with many insightful comments, and even helped me fix typos in the footnotes. I would also like to thank Linda Argote and Geoff Mulgan for reading early drafts of this book and providing me with great comments.

But these acknowledgements would not be complete without thanking some key members of my team: the Center for Collective Learning. Many of them read some versions of this book and provided me with very helpful comments. In particular, I would like to thank Viktor Stojkoski and Philipp Koch, who have become invaluable collaborators in recent years. I also want to thank Lea Karbevska, Mariana Macedo, Melanie Oyarzun, and Jingling Zhang, who read early drafts of this book which we discussed in the court-yard of the Manufacture de Tabacs campus.

But this is also a book that was written in many places. One of them, where I worked religiously on my repeated trips to Budapest, was the Central Café. I would like to especially thank László, a tall and friendly waiter with a passion for basketball. He always found a table for me to park my ass for almost the entire day on Saturdays and Sundays as I transformed eggs Benedict and tea into words.

I would also like to thank the friends and family that read incomplete versions of this book. I would like to particularly thank my cousin Francisca Vallejos, who provided detailed comments on several chapters and entertained discussion about the book's contents. I would also like to thank my cousin Ignacio Vallejos, as well as my friends Adeline Galland and Sonia Ferrer, who read a few chapters of this book during our 2024 summer vacation at the Château de Briat. Finally, I would like to thank Alex Simoes, who provided me with feedback after reading some early versions of the book in the summer of 2023. Alex has been an incredible lifelong partner and collaborator, and I owe him immense gratitude.

I feel very fortunate for everyone who supported me through this journey. Particularly, the great environment I have at home with my wife Anna, our daughter Iris, my mother-in-law Luda, and our chihuahua Toshka.

All mistakes are my own.

Toulouse, January 5, 2025.

# List of Illustrations

# References

1  Bloom, B. S. *Taxonomy of Educational Objectives, Handbook 1: Cognitive Domain* (Addison-Wesley Longman Ltd, London, 1984).

2  Hayek, F. 'The Use of Knowledge in Society,' *The American Economic Review* 35, 519–530 (1945).

3  Foray, D. *The Economics of Knowledge* (MIT Press, 2004).

4  Collins, H. *Tacit and Explicit Knowledge* (University of Chicago Press, 2010).

5  '$602 millones se han invertido en Yachay desde 2013, revela informe de Comisión de Fiscalización sobre cinco universidades emblemáticas del país [$602 million has been invested in Yachay since 2013, as revealed by a report from the Overseeing Commission of five key universities of the country],' *El Universo*. https://www.eluniverso.com/noticias/politica/602-millones-se-han-invertido-en-yachay-desde-2013-revela-informe-de-comision-de-fiscalizacion-sobre-cinco-universidades-emblematicas-del-pais-nota/ (2022).

6  'Según el Gobierno, en proyecto Yachay hubo un despilfarro de alrededor de $1,000 millones [According to the government, the Yachay project wasted around $1,000 million],' *El Universo*. https://www.eluniverso.com/noticias/politica/el-presidente-guillermo-lasso-cuestiono-el-manejo-que-se-habria-dado-al-proyecto-yachay-nota/ (2022).

7  Presidencia de la República del Ecuador Rafael Correa: 'Con Yachay damos un salto hacia la sociedad del conocimiento,' (AUDIO Y VIDEO) [Ecuador's Office of the Presidency of Rafael Correa: 'With Yachay we made a leap into the society of knowledge].' https://www.presidencia.gob.ec/rafael-correa-con-yachay-damos-un-salto-hacia-la-sociedad-del-conocimiento-audio-y-video/.

8  Mack, E. 'Plotting the next Silicon Valley – you'll never guess where,' *CNET*. https://www.cnet.com/news/plotting-the-next-silicon-valley-youll-never-guess-where/.

# References

9 Telégrafo, E. 'Yachay, el salto en la revolución del conocimiento,' (VIDEO). *El Telégrafo*. https://www.eltelegrafo.com.ec/noticias/sociedad/1/yachay-el-salto-en-la-revolucion-del-conocimiento-video (2014).

10 Vega-Villa, K. R. 'Missed opportunities in Yachay,' *Science* 358, 459–459 (2017).

11 Lerner, J. *Boulevard of Broken Dreams: Why Public Efforts to Boost Entre-preneurship and Venture Capital Have Failed – and What to Do about It* (Princeton University Press, Princeton, N.J., 2012).

12 Scheck, J., Jones, R. & Said, S. 'A Prince's $500 Billion Desert Dream: Flying Cars, Robot Dinosaurs and a Giant Artificial Moon,' *Wall Street Journal* (2019).

13 Nereim, Vivian. 'MBS's $500 Billion Desert Dream Just Keeps Get-ting Weirder,' *Bloomberg.com* (2022).

14 Aghion, P. & Howitt, P. 'A Model of Growth through Creative Destruction,' *Econometrica: Journal of the Econometric Society* 323–351 (1992).

15 Lucas, R. E. 'On the Mechanics of Economic Development,' *Journal of Monetary Economics* 22, 3–42 (1988).

16 Audretsch, D. B. & Feldman, M. P. 'R&D Spillovers and the Geog-raphy of Innovation and Production,' *The American Economic Review* 86, 630–640 (1996).

17 Balland, P.-A. *et al.* 'Complex Economic Activities Concentrate in Large Cities,' *Nat Hum Behav* 1–7 (2020) doi:10.1038/s41562-019-0803-3.

18 Argote, L. & Ingram, P. 'Knowledge Transfer: A Basis for Competi-tive Advantage in Firms,' *Organizational Behavior and Human Decision Processes* 82, 150–169 (2000).

19 Argote, L. *Organizational Learning: Creating, Retaining and Transferring Knowledge* (Springer Science & Business Media, 2012).

20 Jaffe, A. B. *Technological Opportunity and Spillovers of R&D: Evidence from Firms' Patents, Profits and Market Value* (National Bureau of Eco-nomic Research, Cambridge, Mass., USA, 1986).

21 Jaffe, A. B., Trajtenberg, M. & Henderson, R. 'Geographic Localiza-tion of Knowledge Spillovers as Evidenced by Patent Citations,' *Q J Econ* 108, 577–598 (1993).

22 Audretsch, D. B. & Feldman, M. P. 'Knowledge Spillovers and the Geography of Innovation,' *Handbook of Regional and Urban Economics* (eds. Henderson, J. V. & Thisse, J.-F.) vol. 4, 2713–2739 (Elsevier, 2004).

23 Isaacson, W. *The Innovators: How a Group of Inventors, Hackers, Geniuses and Geeks Created the Digital Revolution* (Simon and Schuster, 2014).

24 Randolph, M. *That Will Never Work: The Birth of Netflix and the Amazing Life of an Idea* (Endeavour, 2019).

25 'Saudis Scale Back Ambition for $1.5 Trillion Desert Project Neom,' *Bloomberg.com* (2024).

26 Oñate, S. 'Ciudad del Conocimiento Yachay no seguirá funcionando [City of Knowlegde Yachay will cease its functions],' *El Comercio*. https://www.elcomercio.com/actualidad/politica/ciudad-conocimiento-yachay-decreto-lasso.html (2023).

27 'Ejecutivo ordena el cierre de la Ciudad del Conocimiento Yachay [Executive power orders the closing of the City of Knowledge Yachay],' *Primicias*.https://www.primicias.ec/noticias/sociedad/yachay-ciudad-conocimiento-cierre/.

28 'Dr. Theodore Wright, 75, Dies; Leader in Aviation Development,' *The New York Times*. https://www.nytimes.com/1970/08/22/archives/dr-theodore-wright-75-dies-leader-in-aviation-development-194448.html.

29 Wright, T. P. 'Factors Affecting the Cost of Airplanes,' *Journal of the Aeronautical Sciences* 3, 122–128 (1936).

30 Lambert, B. 'Leonard A.Rapping Dies at 57; Adviser and Economic Theorist,' *The New York Times* (1991).

31 Rapping, L. 'Learning and World War II Production Functions,' *The Review of Economics and Statistics* 47, 81–86 (1965).

32 Thompson, P. 'How Much Did the Liberty Shipbuilders Learn? New Evidence for an Old Case Study,' *Journal of Political Economy* 109, 103–137 (2001).

33 Kaplan, J. *et al.* 'Scaling laws for neural language models,' *arXiv preprint arXiv:2001.08361* (2020).

34 Argote, L. & Epple, D. 'Learning Curves in Manufacturing,' *Science* 247, 920–924 (1990).

35  Zhai, X., Kolesnikov, A., Houlsby, N. & Beyer, L. 'Scaling vision transformers,' *Proceedings of the IEEE/CVF Conference on Computer Vision and Pattern Recognition* 12104–12113 (2022).

36  March, J. G. & Simon, H. A. *Organizations* (Wiley-Blackwell, Cambridge, Mass., USA, 1993).

37  Argote, L. & Fahrenkopf, E. 'Knowledge Transfer in Organizations: The Roles of Members, Tasks, Tools, and Networks,' *Organizational Behavior and Human Decision Processes* 136, 146–159 (2016).

38  Thackray, A., Brock, D. C. & Jones, R. *Moore's Law: The Life of Gordon Moore, Silicon Valley's Quiet Revolutionary* (Basic Books, New York, 2015).

39  Moore, G. E. *Cramming More Components onto Integrated Circuits.* (McGraw-Hill, New York, 1965).

40  'Development of the Beckman pH Meter – National Historic Chemical Landmark,' *American Chemical Society.* https://www.acs.org/education/whatischemistry/landmarks/beckman.html.

41  '1948: The European Transistor Invention,' Computer History Museum. https://www.computerhistory.org/siliconengine/the-european-transistor-invention/.

42  'Транзисторная история,' *itWeek.* https://www.itweek.ru/idea/article/detail.php?ID=103634.

43  Nordhaus, W. D. 'Quality Change in Price Indexes,' *Journal of Economic Perspectives* 12, 59–68 (1998).

44  Nordhaus, W. D. 'Do Real-Output and Real-Wage Measures Capture Reality? The History of Lighting Suggests Not,' *The Economics of New Goods* 27–70 (University of Chicago Press, 1996).

45  Fouquet, R. & Pearson, P. J. G. 'Seven Centuries of Energy Services: The Price and Use of Light in the United Kingdom (1300–2000),' *The Energy Journal* 27, 139–177 (2006).

46  'Evolution of solar PV module cost by data source, 1970–2020,' *IEA.* https://www.iea.org/data-and-statistics/charts/evolution-of-solar-pv-module-cost-by-data-source-1970-2020.

47  'DNA Sequencing Costs: Data,' *Genome.gov.* https://www.genome.gov/about-genomics/fact-sheets/DNA-Sequencing-Costs-Data (2022).

48  Bloom, N., Jones, C. I., Van Reenen, J. & Webb, M. 'Are ideas getting harder to find?' *American Economic Review* 110, 1104–1144 (2020).

49  Feynman, R. P. 'There's Plenty of Room at the Bottom,' *Engineering and Science* 23, 22–36 (1960).

50  'Semiconductor Advancement: Is Moore's Law Finally Dead?' *Information Week.* https://www.informationweek.com/it-sectors/semiconductor-advancement-is-moore-s-law-finally-dead-.

51  MacFarquhar, L. 'The Gospel of Disruption,' *The New Yorker* (2012).

52  Henderson, R. M. & Clark, K. B. 'Architectural Innovation: The Reconfiguration of Existing,' *Administrative Science Quarterly* 35, 9–30 (1990).

53  Stone, B. *The Everything Store: Jeff Bezos and the Age of Amazon* (Random House, 2013).

54  Lepore, J. 'What the Gospel of Innovation Gets Wrong,' *The New Yorker* 23, (2014).

55  Christensen, C. M. 'How Will You Measure Your Life?' *Harvard Business Review* (2010).

56  Graeber, T., Roth, C. & Zimmermann, F. 'Stories, Statistics, and Memory,' *The Quarterly Journal of Economics* 139, 2181–2225 (2024).

57  Butler, O. 'Can't Help Falling in Price: Why Elvis Memorabilia is Plummeting in Value,' *The Guardian* (2017).

58  Simon, H. A. 'On a Class of Skew Distribution Functions,' *Biometrika* 42, 425–440 (1955).

59  Price, D. D. S. 'A General Theory of Bibliometric and other Cumulative Advantage Processes,' *J. Am. Soc. Inf. Sci.* 27, 292–306 (1976).

60  Barabási, A.-L. & Albert, R. 'Emergence of Scaling in Random Networks,' *Science* 286, 509–512 (1999).

61  Trust, G. 'Most Billboard Hot 100 Entries among All Acts,' *Billboard.* https://www.billboard.com/photos/most-billboard-hot-100-entries-all-acts/ (2017).

62  Argote, L., Beckman, S. L. & Epple, D. 'The Persistence and Transfer of Learning in Industrial Settings,' *Management Science* 36, 140–154 (1990).

63  'L-1011: Luxury Among the Clouds,' *Lockheed Martin.* https://www.lockheedmartin.com/en-us/news/features/history/l-1011.html.

64  Bonanos, C. *Instant: The Story of Polaroid* vol. 1 (Chronicle Books, 2012).

65  Land, E. H. & Friedman, J. S. 'Polarizing Refracting Bodies,' (1933).

66  Reuters: 'Kodak Settles With Polaroid,' *The New York Times* (1991).

67 'The First Digital Photos,' National Science and Media Museum. https://www.scienceandmediamuseum.org.uk/objects-and-stories/ first-digital-photos.

68 'The Company That's Keeping the Polaroid Legacy Alive,' *Bloomberg.com*.

69 Gampat, C. 'Why is Modern Polaroid Film Nowhere as Good as the Old Stuff?' *The Phoblographer* (2022). https://www.thephoblographer. com/2022/01/24/why-is-modern-polaroid-film-nowhere-as-good-as-the-old-stuff/.

70 *An Impossible Project* (Instant Film, Mischief Films, Egoli Tossell Pictures, 2021).

71 Christakis, N. A. *Blueprint: The Evolutionary Origins of a Good Society* (Hachette UK, 2019).

72 'Project Chess: The Story Behind the Original IBM PC,' *PCMAG* https://www.pcmag.com/news/project-chess-the-story-behind-the-original-ibm-pc.

73 Bagnall, W. R. *Samuel Slater and the Early Development of the Cotton Manufacture in the United States* (J. S. Stewart, Printer and Bookbinder, 1890).

74 Goodrich, M. 'Pawtucket and the Slater Centennial,' *New England Magazine* 3, 138 (1890).

75 Griliches, Z. 'Hybrid Corn: An Exploration in the Economics of Technological Change,' *Econometrica, Journal of the Econometric Society* 501–522 (1957).

76 Krugman, P. *Geography and Trade* (MIT Press, 1992).

77 Jaffe, A. B. 'Real Effects of Academic Research,' *American Economic Review* 79, 957–970 (1989).

78 Barabási, A. *Linked: How Everything Is Connected to Everything Else and What It Means for Business, Science, and Everyday Life* (Basic Books, New York, 2014).

79 Watts, D. J. *Six Degrees: The Science of a Connected Age* (W. W. Norton & Company, 2004).

80 Agrawal, A., Cockburn, I. & McHale, J. 'Gone But Not Forgotten: Knowledge Flows, Labor Mobility, and Enduring Social Relationships,' *J Econ Geogr* 6, 571–591 (2006).

81 Breschi, S. & Lissoni, F. 'Knowledge Networks from Patent Data,' *Handbook of Quantitative Science and Technology Research* 613–643 (Springer, 2004).

82 Balconi, M., Breschi, S. & Lissoni, F. 'Networks of Inventors and the Role of Academia: An Exploration of Italian Patent Data,' *Research Policy* 33, 127–145 (2004).

83 Singh, J. 'Collaborative Networks as Determinants of Knowledge Diffusion Patterns,' *Management Science* 51, 756–770 (2005).

84 Hidalgo, C. *Why Information Grows: The Evolution of Order, from Atoms to Economies* (Basic Books, New York, 2015).

85 Granovetter, M. S. *Getting a Job: A Study of Contacts and Careers* (Harvard University Press, Cambridge, Mass, 1974).

86 Granovetter, M. S. 'The Strength of Weak Ties,' *American Journal of Sociology* 78, 1360–1380 (1973).

87 Granovetter, M. 'Economic Action and Social Structure: The Problem of Embeddedness,' *American Journal of Sociology* 91, 481–510 (1985).

88 Moser, P., Voena, A. & Waldinger, F. 'German Jewish Émigrés and US Invention,' *American Economic Review* 104, 3222–3255 (2014).

89 Waldinger, F. 'Peer Effects in Science: Evidence from the Dismissal of Scientists in Nazi Germany,' *The Review of Economic Studies* 79, 838–861 (2012).

90 Waldinger, F. 'Quality Matters: The Expulsion of Professors and the Consequences for PhD Student Outcomes in Nazi Germany,' *Journal of Political Economy* 118, 787–831 (2010).

91 Ganguli, I. 'Immigration and Ideas: What Did Russian Scientists 'Bring' to the United States?' *Journal of Labor Economics* 33, S257–S288 (2015).

92 Borjas, G. J. & Doran, K. B. 'Cognitive Mobility: Labor Market Responses to Supply Shocks in the Space of Ideas,' *Journal of Labor Economics* 33, S109–S145 (2015).

93 Borowiecki, K. J. 'Good Reverberations? Teacher Influence in Music Composition since 1450,' *Journal of Political Economy* 130, 991–1090 (2022).

94 Henrich, J. *The Secret of Our Success: How Culture Is Driving Human Evolution, Domesticating Our Species, and Making Us Smarter* (Princeton University Press, Princeton, 2015).

95 Chudek, M., Heller, S., Birch, S. & Henrich, J. 'Prestige-Biased Cultural Learning: Bystander's Differential Attention to Potential Models Influences Children's Learning,' *Evolution and Human Behavior* 33, 46–56 (2012).

96 Ryalls, B. O., Gul, R. E. & Ryalls, K. R. 'Infant Imitation of Peer and Adult Models: Evidence for a Peer Model Advantage,' *Merrill-Palmer Quarterly* 46, 188–202 (2000).

97 Fairlie, R. W., Hoffmann, F. & Oreopoulos, P. 'A Community College Instructor like Me: Race and Ethnicity Interactions in the Classroom,' *American Economic Review* 104, 2567–2591 (2014).

98 Azoulay, P., Liu, C. C. & Stuart, T. E. 'Social Influence Given (Partially) Deliberate Matching: Career Imprints in the Creation of Academic Entrepreneurs,' *American Journal of Sociology* 122, 1223–1271 (2017).

99 Collins, H. M. 'The TEA Set: Tacit Knowledge and Scientific Networks,' *Science Studies* 4, 165–185 (1974).

100 Parsons, C. & Vézina, P.-L. 'Migrant Networks and Trade: The Vietnamese Boat People as a Natural Experiment,' *Economic Journal* (2017).

101 Scoville, W. C. 'The Huguenots and the Diffusion of Technology. I,' *Journal of Political Economy* 60, 294–311 (1952).

102 Scoville, W. C. 'The Huguenots and the Diffusion of Technology. II,' *Journal of Political Economy* 60, 392–411 (1952).

103 Cohen, W. M. & Levinthal, D. A. 'Absorptive Capacity: A New Perspective on Learning and Innovation,' *Administrative Science Quarterly* 35, 128–152 (1990).

104 Cohen, W. M. & Levinthal, D. A. 'Innovation and Learning: The Two Faces of R&D,' *The Economic Journal* 99, 569–596 (1989).

105 Kerr, W. R. *The Gift of Global Talent: How Migration Shapes Business, Economy & Society* (Stanford University Press, 2018).

106 Lissoni, F. & Miguelez, E. 'Migration and Innovation: Learning from Patent and Inventor Data,' *Journal of Economic Perspectives* 38, 27–54 (2024).

107 Balland, P.-A. & Rigby, D. 'The Geography of Complex Knowledge,' *Economic Geography* 93, 1–23 (2017).

108  Weaver, W. 'Science and Complexity,' *American Scientist* 36, 536–544 (1948).

109  Hidalgo, C. A. 'Economic Complexity Theory and Applications,' *Nature Reviews Physics* 1–22 (2021).

110  Autant-Bernard, C. 'Science and Knowledge Flows: Evidence From the French Case,' *Research policy* 30, 1069–1078 (2001).

111  West, P. &. 'Godfather of the Modern Scooter | Corradino D'Ascanio,' *Visordown*, April 26, 2021. https://www.visordown.com/features/general/godfather-modern-scooter-corradino-d%E2%80%99ascanio.

112  *WideMagazine*. https://wide.piaggiogroup.com/en/articles/people/corradino-d-ascanio-true-genius/index.html.

113  JWH1975. 'What Happened to Japan's WWII Aircraft Companies after 1945,' *wwiiafterwwii* (2022). https://wwiiafterwwii.wordpress.com/2022/10/21/what-happened-to-japans-wwii-aircraft-companies-after-1945/.

114  Narahashi, I. 'Some Aspects of the New Civil Aeronautics Law of Japan,' *Journal of Air Law and Commerce*, 24:1.

115  'The Jet Engine: A Historical Introduction.' https://cs.stanford.edu/people/eroberts/courses/ww2/projects/jet-airplanes/planes.html.

116  'Heinkel Aircraft Works.' https://www.centennialofflight.net/essay/Aerospace/Heinkel/Aero57.htm.

117  Neffke, F., Henning, M. & Boschma, R. 'How Do Regions Diversify over Time? Industry Relatedness and the Development of New Growth Paths in Regions,' *Economic Geography* 87, 237–265 (2011).

118  Kogler, D. F., Rigby, D. L. & Tucker, I. 'Mapping Knowledge Space and Technological Relatedness in US Cities,' *Global and Regional Dynamics in Knowledge Flows and Innovation* 58–75 (Routledge, 2015).

119  Boschma, R., Balland, P.-A. & Kogler, D. F. 'Relatedness and Technological Change in Cities: The Rise and Fall of Technological Knowledge in US Metropolitan Areas From 1981 To 2010,' *Ind Corp Change* 24, 223–250 (2015).

120  Guevara, M. R., Hartmann, D., Aristarán, M., Mendoza, M. & Hidalgo, C. A. 'The Research Space: Using Career Paths to Predict the Evolution of the Research Output of Individuals, Institutions, and Nations,' *Scientometrics* 109, 1695–1709 (2016).

121  Hidalgo *et al.* 'The Principle of Relatedness,' *Unifying Themes in Complex Systems IX* (eds. Morales, A. J., Gershenson, C., Braha, D., Minai, A. A. & Bar-Yam, Y.), 451–457 (Springer International Publishing, 2018).

122  Ellison, G. & Glaeser, E. L. 'Geographic Concentration in US Manufacturing Industries: A Dartboard Approach,' *Journal of Political Economy* 105, 889–927 (1997).

123  Ellison, G., Glaeser, E. L. & Kerr, W. R. 'What Causes Industry Agglomeration? Evidence from Coagglomeration Patterns,' *American Economic Review* 100, 1195–1213 (2010).

124  Diodato, D., Neffke, F. & O'Clery, N. 'Why Do Industries Coagglomerate? How Marshallian Externalities Differ by Industry and Have Evolved over Time,' *Journal of Urban Economics* 106, 1–26 (2018).

125  Jara-Figueroa, C., Jun, B., Glaeser, E. L. & Hidalgo, C. A. 'The Role of Industry-Specific, Occupation-Specific, and Location-Specific Knowledge in the Growth and Survival of New Firms,' *PNAS* 115, 12646–12653 (2018).

126  Jun, B., Alshamsi, A., Gao, J. & Hidalgo, C. A. 'Bilateral Relatedness: Knowledge Diffusion and The Evolution of Bilateral Trade', *Journal of Evolutionary Economics* 1–31 (2019).

127  Miguelez, E. & Morrison, A. 'Migrant Inventors as Agents of Technological Change,' *J Technol Transf* 48, 669–692 (2023).

128  Di Iasio, V. & Miguelez, E. 'The Ties That Bind and Transform: Knowledge Remittances, Relatedness and the Direction of Technical Change,' *Journal of Economic Geography* 22, 423–448 (2022).

129  'John J. McCloy, Lawyer and Diplomat, Is Dead at 93,' *The New York Times* (1989).

130  Congressional Record. https://www.congress.gov/bound-congressional-record/1948/03/04/94/senate-section/article/2095-2135.

131  'Data on France | Reconstructing Global Inequality'. https://clio-infra.eu/Countries/France.html.

132  Andrews, M. *The Limits of Institutional Reform in Development: Changing Rules for Realistic Solutions* (Cambridge University Press, 2013).

133  North, D. C. *Institutions, Institutional Change and Economic Performance* (Cambridge University Press, 1990).

134  Fukuyama, F. 'The End of History?' *The National Interest* 3–18 (1989).

135 The Sveriges Riksbank Prize in Economic Sciences in Memory of Alfred Nobel 1993. *NobelPrize.org*. https://www.nobelprize.org/prizes/economic-sciences/1993/north/facts/.

136 Williamson, J. 'A Short History of the Washington Consensus,' *Law & Bus. Rev. Am.* 15, 7 (2009).

137 Rodrik, D., Subramanian, A. & Trebbi, F. 'Institutions Rule: The Primacy of Institutions over Geography and Integration in Economic Development,' *Journal of Economic Growth* 9, 131–165 (2004).

138 Rodrik, D. 'Institutions, Integration, and Geography: In Search of the Deep Determinants of Economic Growth,' *Search of Prosperity: Analytic Country Studies on Growth* 1–30 (2003).

139 Acemoglu, D., Johnson, S. & Robinson, J. A. 'The Colonial Origins of Comparative Development: An Empirical Investigation,' *American Economic Review* 91, 1369–1401 (2001).

140 Glaeser, E. L., La Porta, R., Lopez de Silanes, F. & Shleifer, A. 'Do Institutions Cause Growth?' *Journal of Economic Growth* 9, 271–303 (2004).

141 Rodrik, D. 'What's So Special about China's Exports?' *China & World Economy* 14, 1–19 (2006).

142 Ken, N. *Zhong Guan Village: Tales from the Heart of China's Silicon Valley* (ACA Publishing Limited, 2022).

143 Weber, I. M. *How China Escaped Shock Therapy: The Market Reform Debate* (Routledge, London, 2021). doi:10.4324/9780429490125.

144 Ang, Y. Y. *How China Escaped the Poverty Trap* (Cornell University Press, 2016).

145 McCaskey, J. P. 'History of 'Temperature': Maturation of a Measurement Concept,' *Annals of Science* 77, 399–444 (2020).

146 Hausmann, R., Hwang, J. & Rodrik, D. 'What You Export Matters,' *J Econ Growth* 12, 1–25 (2007).

147 Hidalgo, C. A. & Hausmann, R. 'The Building Blocks of Economic Complexity,' *PNAS* 106, 10570–10575 (2009).

148 Mikolov, T., Chen, K., Corrado, G. & Dean, J. 'Efficient Estimation of Word Representations in Vector Space,' *arXiv preprint arXiv:1301.3781* (2013).

149 Servedio, V. D., Bellina, A., Calò, E. & De Marzo, G. 'Economic Complexity in Mono-Partite Networks,' *arXiv preprint arXiv:2405.04158* (2024).

150 Hartmann, D., Guevara, M. R., Jara-Figueroa, C., Aristarán, M. & Hidalgo, C. A. 'Linking Economic Complexity, Institutions, and Income Inequality,' *World Development* 93, 75–93 (2017).

151 Ben Saâd, M. & Assoumou-Ella, G. 'Economic Complexity and Gender Inequality in Education: An Empirical Study,' *Economics Bulletin* 39, 321–334 (2019).

152 Chu, L. K. & Hoang, D. P. 'How Does Economic Complexity Influence Income Inequality? New Evidence from International Data,' *Economic Analysis and Policy* 68, 44–57 (2020).

153 Lee, K.-K. & Vu, T. V. 'Economic Complexity, Human Capital and Income Inequality: A Cross-Country Analysis,' *The Japanese Economic Review* 1–24 (2019).

154 Fawaz, F. & Rahnama-Moghadamm, M. 'Spatial Dependence of Global Income Inequality: The Role of Economic Complexity,' *The International Trade Journal* 33, 542–554 (2019).

155 Hartmann, D. & Pinheiro, F. L. *Economic Complexity and Inequality at the National and Regional Level.* http://arxiv.org/abs/2206.00818 (2022) doi:10.48550/arXiv.2206.00818.

156 'The life of Meroë Marston Morse and Her Polaroid legacy,' *History* https://www.nationalgeographic.com/history/article/polaroid-meroe-marston-morse-photography (2024).

157 McElheny, V. K. *Insisting on the Impossible: The Life of Edwin Land* (Basic Books, Cambridge, Mass., 1999).

158 Romero, J. P. & Gramkow, C. 'Economic Complexity and Greenhouse Gas Emissions,' *World Development* 139, 105317 (2021).

159 Lapatinas, A., Garas, A., Boleti, E. & Kyriakou, A. 'Economic Complexity and Environmental Performance: Evidence From a World Sample,' (2019).

160 Chu, L. K. 'Economic Structure and Environmental Kuznets Curve Hypothesis: New Evidence From Economic Complexity,' *Applied Economics Letters* 1–5 (2020).

161 Neagu, O. 'The Link between Economic Complexity and Carbon Emissions in the European Union Countries: A Model Based on the Environmental Kuznets Curve (EKC) Approach,' *Sustainability* 11, 4753 (2019).

162 Gozgor, G. & Can, M. 'Export Product Diversification and the Environmental Kuznets Curve: Evidence From Turkey,' *Environmental Science and Pollution Research* 23, 21594–21603 (2016).

163 Montiel-Hernández, M. G., Pérez-Hernández, C. C. & Salazar-Hernández, B. C. 'The Intrinsic Links of Economic Complexity with Sustainability Dimensions: A Systematic Review and Agenda for Future Research,' *Sustainability* 16, 391 (2024).

164 Chávez, J. C., Mosqueda, M. T. & Gómez-Zaldívar, M. 'Economic Complexity and Regional Growth Performance: Evidence from the Mexican Economy,' *Review of Regional Studies* 47, 201–219 (2017).

165 Fritz, B. S. & Manduca, R. A. 'The Economic Complexity of US Metropolitan Areas,' *Regional Studies* 1–12 (2021).

166 Stojkoski, V., Koch, P. & Hidalgo, C. A. 'Multidimensional Economic Complexity and Inclusive Green Growth,' *Commun Earth Environ* 4, 1–12 (2023).

167 Li, X., Linde, S. & Shimao, H. 'Major Complexity Index and College Skill Production,' SSRN Scholarly Paper. https://doi.org/10.2139/ssrn.3791651 (2023).

168 Juhász, S., Wachs, J., Kaminski, J. & Hidalgo, C. A. 'The Software Complexity of Nations.' Preprint. https://doi.org/10.48550/arXiv.2407.13880 (2024).

169 Vallim, A. & Monasterio, L. 'Crescimento econômico regional e complexidade: O papel das microempresas e dos MEI,' *Revista Brasileira de Estudos Regionais e Urbanos* 17, 131–154 (2023).

170 Balsalobre, S. J. P., Verduras, C. L. & Lanchas, J. D. 'Measuring the Economic Complexity at the Sub-National Level Using International and Interregional Trade,' *Nineteenth Annual Conference of European Trade Study Group* (2017).

171 Stojkoski, V., Utkovski, Z. & Kocarev, L. 'The Impact of Services on Economic Complexity: Service Sophistication as Route for Economic Growth,' *PLOS ONE* 11, e0161633 (2016).

172 Ourens, G. 'Can the Method of Reflections Help Predict Future Growth?' *Documento de Trabajo/FCS-DE; 17/12* (2012).

173 Domini, G. 'Patterns of Specialization and Economic Complexity

through the Lens of Universal Exhibitions, 1855–1900,' *Explorations in Economic History* 83, 101421 (2022).

174 Poncet, S. & de Waldemar, F. S. 'Economic Complexity and Growth,' *Revue économique* 64, 495–503 (2013).

175 Salinas, G. 'Proximity and Horizontal Policies: The Backbone of Export Diversification,' *IMF.* https://www.imf.org/en/Publications/WP/Issues/2021/03/05/Proximity-and-Horizontal-Policies-The-Backbone-of-Export-Diversification-50141.

176 Ministry of Investment Trade and Industry. 'New Industrial Master Plan 2030.' (2023).

177 Richerson, P. J. & Boyd, R. *Not by Genes Alone: How Culture Transformed Human Evolution* (University of Chicago Press, Chicago, 2004).

178 Acemoglu, D. & Johnson, S. *Power and Progress: Our Thousand-Year Struggle over Technology and Prosperity* (PublicAffairs, 2023).

179 Pettegree, A. *The Book in the Renaissance* (Yale University Press, 2010).

180 Jara-Figueroa, C., Yu, A. Z. & Hidalgo, C. A. 'How the Medium Shapes the Message: Printing and the Rise of the Arts and Sciences,' *PLOS ONE* 14, e0205771 (2019).

181 McLuhan, M. *Understanding Media: The Extensions of Man* (Gingko Press, 2003).

182 Lee, K.-F. *AI Superpowers: China, Silicon Valley, and the New World Order* (Houghton Mifflin Harcourt, 2018).

183 Lee, K. *The Art of Economic Catch-Up: Barriers, Detours and Leapfrogging in Innovation Systems* (Cambridge University Press, Cambridge, 2019). doi:10.1017/9781108588232.

184 Zuckerman, G. *A Shot to Save the World: The Inside Story of the Life-or-Death Race for a COVID-19 Vaccine* (Portfolio, New York, 2021).

185 Diamond, J. M. *Guns, Germs, and Steel: The Fates of Human Societies.* (W. W. Norton & Company, New York, 1999).

186 Seabright, P. 'Skill versus Judgement and the Architecture of Organisations,' *European Economic Review* 44, 856–868 (2000).

187 Lee, K. 'Singapore's Founding Father Thought Air Conditioning Was the Secret to His Country's Success,' *Vox.* https://www.vox.com/2015/3/23/8278085/singapore-lee-kuan-yew-air-conditioning (2015).

188 'Lee Kuan Yew, Truly the Father of Changi Airport,' *The Business Times*. https://www.businesstimes.com.sg/opinion-features/columns/lee-kuan-yew-truly-father-changi-airport (2015).

189 'Lee Kuan Yew – Charlie Rose Interview' (18th October 2000).

190 'Historia | Universidad San Francisco de Quito.' https://www.usfq.edu.ec/es/historia.

191 *The Discovery That Transformed Pi* (2021).

192 Hidalgo, C. A. 'The Complexity of Increasing Returns, *The Bridge* 50, 29–30 (2020).

193 Arthur, W. B. 'Positive Feedbacks in the Economy,' *Scientific American* 262, 92–99 (1990).

194 Arthur, W. B. 'Increasing Returns and the New World of Business,' *Harvard Business Review* (1996).

195 Lee, K. & Malerba, F. 'Catch-up Cycles and Changes in Industrial Leadership: Windows of Opportunity and Responses of Firms and Countries in the Evolution of Sectoral Systems,' *Research Policy* 46, 338–351 (2017).

196 'Executive summary – Batteries and Secure Energy Transitions – Analysis.' *IEA*. https://www.iea.org/reports/batteries-and-secure-energy-transitions/executive-summary.

197 'Electricity – Renewables 2023 – Analysis.' *IEA*. https://www.iea.org/reports/renewables-2023/electricity.

198 'Nuclear.' *IEA*. https://www.iea.org/energy-system/electricity/nuclear-power.

199 Westwood, J. N. 'John Hughes and Russian Metallurgy,' *The Economic History Review* 17, 564–569 (1965).

200 Bahar, D., Rosenow, S., Stein, E. & Wagner, R. 'Export Take-offs and Acceleration: Unpacking Cross-Sector Linkages in the Evolution of Comparative Advantage,' *World Development* 117, 48–60 (2019).

201 Karbevska, L. & Hidalgo, C. A. 'Mapping Global Value Chains at the Product Level,' *arXiv preprint arXiv:2308.02491* (2023).

202 Alshamsi, A., Pinheiro, F. L. & Hidalgo, C. A. 'Optimal Diversification Strategies in the Networks of Related Products and of Related Research Areas,' *Nature Communications* 9, 1328 (2018).

203 Hidalgo, C. A. 'The Policy Implications of Economic Complexity,' *Research Policy* 52, 104863 (2023).

204 Yañez, L. *Pepo es de Conce* (Trama Impresores, S.A., Hualpen, Bio Bio, Chile, 2020).

205 Hausmann, R. *et al. The Atlas of Economic Complexity: Mapping Paths to Prosperity* (MIT Press, 2014).

206 Wheeler, J. A. 'Information, Physics, Quantum: The Search for Links,' *Proceedings III International Symposium on Foundations of Quantum Mechanics* (ed. Archibald, W. J.) 354–358 (1989).

207 Schrödinger/Penrose, E. *What Is Life?: With Mind and Matter and Autobiographical Sketches* (Cambridge University Press, Cambridge, 2012).

# Index